Being Human

Being Human

HOW OUR BIOLOGY SHAPED WORLD HISTORY

LEWIS DARTNELL

THE BODLEY HEAD
LONDON

1 3 5 7 9 10 8 6 4 2

The Bodley Head, an imprint of Vintage, is part of the Penguin
Random House group of companies whose addresses can be found
at global.penguinrandomhouse.com

Penguin
Random House
UK

First published by The Bodley Head in 2023

penguin.co.uk/vintage

A CIP catalogue record for this book is available from the British Library

HB ISBN 9781847926708
TPB ISBN 9781847926715

Typeset in 11.5/14pt Sabon LT Std by Jouve (UK), Milton Keynes
Printed and bound in Great Britain by Clays Ltd, Elcograf S.p.A.

The authorised representative in the EEA is Penguin Random House Ireland,
Morrison Chambers, 32 Nassau Street, Dublin DO2 YH68

Penguin Random House is committed to a sustainable future
for our business, our readers and our planet. This book is made
from Forest Stewardship Council® certified paper.

For Davina & Sebastian

Contents

Introduction

History makes no sense without prehistory, and prehistory makes no sense without biology.
—Edward O. Wilson, *The Social Conquest of Earth*

Humans are an exquisitely intelligent and capable species of ape. Not only are our complex brains a wonder of evolution, but our bodies are engineering marvels. Our physiology has been fine-tuned for efficient long-distance running; our hands are elegantly dextrous for manipulating and making; and our throats and mouths give us astonishing control over the sounds we make. We are virtuoso communicators, with myriad forms of language, able to convey everything from physical instructions to abstract concepts, and to coordinate ourselves in teams and communities. We learn from each other, from our parents and peers, so new generations don't have to start from scratch. Our culture is cumulative: we have amassed our capabilities over time. We have progressed from master crafters of stone tools to wielders of technologies such as supercomputers and spacecraft.

But we're also deeply flawed, both physically and mentally. In many ways, humans just don't work particularly well.

What do US presidents George W. Bush and Ronald Reagan have in common with actors Elizabeth Taylor and Halle Berry? They all almost choked to death on their food (a pretzel, peanut, chicken bone and fig, respectively).[1] In fact, choking is the third-leading cause of death at home today.[2] Compared to any other animal, we seem breathtakingly inept (literally) at the key survival skill of eating without accidentally killing ourselves. The reason for this relates to the changes to our throat that enabled us to form the complex sounds of speech and so become such expressive vocal communicators. During the evolution of our species, the larynx rose higher in the neck and changed its structure to allow more control over sound generation. In all mammals, the pipelines for breathing and eating share a short section of the same tube, with a small flap called the epiglottis serving as a trapdoor to close off the windpipe when swallowing. But the remodelling of the human throat significantly increased the chances of food getting stuck in the windpipe.[3] As Darwin noted: 'Every particle of food and drink which we swallow has to pass over the orifice of the trachea, with some risk of falling into the lungs.'[4]

This is only one of a number of basic design flaws in the architecture of the human body. We evolved to walk upright, but the posture puts huge strain on our knees, and back pain strikes most of us in our lifetimes. The wrist and ankle joints contain pointless vestigial bones that restrict movement and render us susceptible to twists and sprains.[5] We have a number of nerves that take ludicrously long and indirect routes through the body, as well as muscles that no longer serve any purpose (such as those used by other animals to twist their ears). The light-sensitive layer at the back of our eye – the retina – faces back-to-front, leaving us with blind spots in our vision. We're also riddled with defects in our biochemistry and DNA – data-corrupted genes that no longer work – which means, for instance, that we must eat a diet more varied than almost any other animal to obtain the nutrients we need to survive. And our brains, far from being perfectly rational thinking machines, are full of cognitive glitches and bugs. We're also prone to addictions that drive compulsive behaviour, sometimes along self-destructive paths.

Many of these apparent faults are the result of evolutionary compromise. When a particular gene or anatomical structure is needed to satisfy several conflicting demands at the same time, no one function can be perfectly optimised. Our throats must be suitable not only for breathing and eating, but also for articulating speech. Our brains need to make survival decisions in complex, unpredictable environments, but they need to do so with incomplete information and, crucially, very rapidly. It is clear that evolution strives not for the perfect, but merely the good-enough.

What's more, evolution is restricted, in finding solutions to new conditions and survival problems, to tinkering with what is already at its disposal. It never gets the chance to go back to the drawing board and redesign from scratch. We have emerged from our evolutionary history as a palimpsest of overlaying designs, with each new adaptation modifying, or being built on top of, what already existed. Our spine, for instance, is poorly conceived to support an upright posture with a large head on top, but we had to make do with the backbone handed down to us from our ancestors who walked on all fours.

To be human is to be the sum total of all our capabilities and constraints – both our flaws and our faculties make us who we are. And the story of human history has played out in the balance between them.

We migrated from our evolutionary cradle in Africa to become the most widely distributed terrestrial animal species on the planet. Around ten millennia ago, we learned to domesticate wild plants and animals to invent agriculture, and out of this grew increasingly complex social organisations: cities, civilisations, empires. And over this whole, staggering breadth of time, through growth and stagnation, progress and regression, cooperation and conflict, slavery and emancipation, trading and raiding, invasions and revolutions, plagues and wars – through all this tumult and fervour, there has been one constant: ourselves. In almost all key aspects of our physiology and psychology, we're basically the same as our ancestors living in Africa

100,000 years ago. Across cultures worldwide there's a wonderful diversity of beliefs, practices and customs, but while there are superficial differences in our appearances, and more significant genetic variations, to all intents and purposes we are built identically. The fundamental aspects of what it means to be human – the hardware of our bodies and the software of our minds – haven't changed.

In this book, I want to take a deep dive into human history and explore how our fundamental humanness has expressed itself in our cultures, societies and civilisations. How have different quirks of our genetics, biochemistry, anatomy, physiology and psychology manifested themselves, and what have been the consequences and ramifications – not just in terms of singular, momentous events but for the over-arching constants and long-term trends of world history?

As well as the idiosyncrasies of our humanity, we'll explore what we share in our body and behaviour with other animals. Much of our refined culture and society is no more than a thin veneer over our inherent animal nature. We are often no different from other beasts when it comes down to competing for resources and sex or trying to give our children the best chances in life. These primal drives have manifested themselves through history in everything from our family structures to the efforts of royal dynasties to control their bloodlines. We'll take in the latest research in anthropology and sociology, and also see how many aspects of our everyday lives are deeply rooted in our biology.

Many of the requirements and restrictions of our bodies are obvious. We can survive within only a certain range of temperatures, and the efficiency with which our lungs can extract oxygen from the air limits how high we can live. (The highest permanent settlement today is the town of La Rinconada, at 5,100 metres of altitude in the Peruvian Andes.) Our need for a constant intake of water and nutrients to survive also determines the environments in which we can permanently settle. Our inability to drink seawater has historically limited oceanic voyages that relied on supplies of fresh water. Our life cycle, with the long period of development before reaching sexual maturity, governs how quickly we can reproduce and grow populations. Our bodies are vulnerable to being invaded by microscopic organisms and other parasites, which can have fatal consequences. The force our muscles can exert limits the achievements of our labour and has driven us both to harness beasts of burden such as the ox, camel and horse and to develop technology. And our need to sleep dictates the activity cycles of society.[6]

But features of our body have influenced human cultural development – the customs, behaviour and skills that we learn from each other – in more subtle ways as well.

All spoken languages use intricate sequences of sounds created by our upper respiratory tract: air is exhaled from the lungs, and the vibrations of the vocal cords are modified by our throat, mouth, tongue and lips. This sophisticated capability for vocalisation is considered one of the defining characteristics of our species.

Speech is composed of a series of open sounds or vowels – such as ah, ee, oo – interspersed with a greater diversity of consonants: collectively, these are the phonemes of language. Consonant phonemes can be created in a large variety of ways: the plosive release of air for a 'p' or 't'; the fricative restriction of airflow within the mouth for 'f' or 's'; the steady airflow around the sides of the tongue for 'l'; the nasal resonance of 'n'. All the world's languages are composed of a total inventory of around 90 different sounds, although most don't use more than about half of them;[7] English, for example, is composed of around 44 discrete phonemes.[8] By far the most common consonant sound in human speech is 'm', which seems to be the simplest to form. It's used in 95 per cent of the 450 languages studied in detail by the UCLA Phonological Segment Inventory Database (UPSID) – from Abipon to Zuni (and including !Xu).[9] This widespread phoneme is produced by bringing both lips together and sending air through the nose, and it is similar to the lip-smack behaviour of chimpanzees and other primates.[10] It's the phoneme that begins one of the first words ever spoken by over 5 billion of us: a linguistic variant of 'mama'. Thus all languages around the world are dominated by the sounds we find easiest to produce – by the anatomical limits of being human.

Some features of our bodies have profoundly influenced not only what we're physically capable of but how we think about the world. The fact that we have five fingers on each hand (and five toes on each foot) – that we are *pentadactyl* – is an evolutionary happenstance that became fixed in our fish-like ancestors around 350 million years ago. (It is shared by all other four-limbed vertebrates, from crocodiles to birds to dolphins.) But it has come to have profound implications for our conception of numbers and numerical calculation. We have ten digits to count on, and so most ancient cultures around the world adopted a base-ten numerical system.* We think in round numbers of tens or hundreds or thousands – rather than in multiples of, say, six, 36 and 216, as we might if we were tridactyl. By the fifth century AD, the Indo-Arabic numeral system had devised the place-value notation which then developed into modern decimal numbers and the metric system for measurements. Our entire conception of mathematics is ultimately founded on the number of digits that sprout off our forelimbs.

Other aspects of the world we created are inextricably linked to our anatomical traits too. The beat of the second is roughly equivalent to our resting heart rate; the inch was traditionally the thickness of a thumb; and the mile was defined as a thousand Roman paces and thus the composite of our base-ten counting system and the length of the leg.

As we'll see, it's not just our physical features that have left indelible marks on our world. Our evolved psychological

---

* There were exceptions, however. The ancient Sumerians, for example, used a combination of base-ten and base-sixty (which is useful as lots of numbers divide into it), which is the reason why we still split an hour of time into sixty minutes and then sixty seconds, and why there are 360 degrees in a circle.

mechanisms and predispositions have influenced human culture in very particular ways. Many of these are so deeply ingrained in everyday life that we tend to overlook their biological roots. For example, we have a strong tendency towards herd behaviour – fitting in with those in our community by copying their decisions. In evolutionary terms this has served us well. In the natural world full of dangers, it is probably safer to follow everyone else, even if you're not convinced it's the best course of action, rather than risk going it alone. Often, even if we feel we're right, we are loath to stand out from the pack. Such herd behaviour is a way of crowd-sourcing information – others may know something we don't – and can serve as a quick judgement tool, allowing us to economise on the time and cognitive effort in deciding everything for ourselves from scratch. For example, walking through an unfamiliar city looking for a good place for dinner, we're naturally drawn to the busy restaurant rather than the empty one next door.

This herding bias has caused the surges of fads and fashions throughout history. It influences the adoption of other cultural norms, religious views or political preferences as well. But the same psychological bias also destabilises markets and financial systems. The dot-com boom of the late 1990s, for instance, was driven by investors piling in to back internet companies even though many of the start-ups were not financially sound. Investors followed one another, assuming that others had a more reliable assessment or simply not wanting to be left behind in the frenzy, only for the bubble to burst and stock markets to fall sharply after early 2000. Such speculative bubbles have recurred through history since 'tulip mania' in the early-seventeenth-century Netherlands, and the same herding behaviour is behind modern boom and bust cycles such as in cryptocurrency markets.

This book is the third in a trilogy of titles – each of which can be read separately – in which I wanted to explore the grand scale of history and the making of our modern world from a different angle. The first was *The Knowledge: How to Rebuild*

Our World from Scratch, which used the conceit of a manual on how to reboot civilisation as quickly as possible after some kind of apocalypse. It used the notion of the loss of everything that we take for granted today to peer behind the scenes of the modern world, explore how it all works and reveal how different discoveries and inventions enabled humanity's progress. The second book, *Origins: How the Earth Shaped Human History*, zoomed out and explored how features of the planet we live on – from plate tectonics to climate belts, from mineral resources to atmospheric circulation – have profoundly influenced the human story. *Origins* took us from the emergence of our species in the giant crack of the East African rift valley, through millennia of rising and falling civilisations and empires, right into the modern world, showing how the distinct fingerprint of the natural world can be discerned even in politics today.

What I want to do in this book is extend this line of inquiry and put the focus on us – to tell the human story from the perspective of biology and the essence of what it means to be human. I am a biologist by training, and so for me this represents something of a return to my home turf. I'm hoping to reveal the profound and often surprising ways in which intrinsic aspects of our anatomy, genetics, biochemistry and psychology have left their mark on human history.

We'll explore how romantic love and the human family developed as a consequence of our quirky evolution, and how marriage came to be exploited as a political tool by ruling dynasties. Why were European royal families particularly prone to unreliable reproduction, and how did other dynasties solve the problem – in the process creating sterile soldiers akin to those of ant colonies?

We'll take a detailed look at how our vulnerability to infectious diseases has played a multitude of pivotal roles in the history of the world. How did endemic diseases lead to the political union of England and Scotland or help double the size of the United States overnight? Epidemics helped the spread of a once-obscure religion and ushered in the decline of feudalism but also drove the transatlantic African slave trade to the Americas.

Fundamental features of human populations such as growth rate and the balance of males and females can have far-reaching consequences, and we'll explore the effects of such demographic forces. We'll also discover ways to alter our state of consciousness, and how by affecting our minds psychoactive substances came to change the world. We'll explore how alcohol became an intoxicating social lubricant, the stimulating impact of tea and coffee, the invigorating moreishness of tobacco, and how the poppy was wielded as a tool of imperial subjugation.

Errors in our genetic code have far-reaching ramifications. We'll see how a rare mutation that arose in Queen Victoria had disastrous consequences for the ruling dynasties across Europe a century later and also had a hand in the Russian Revolution. Another defunct gene, shared by all of humanity, played a defining role during the Age of Sail and inadvertently led to the emergence of the world's most notorious criminal organisation.

Finally, we'll explore the wide-ranging consequences of bugs in our mental software. Which particular cognitive bias gripped Columbus, was a powerful factor that led to the invasion of Iraq half a millennium later, and today lurks behind the problem of political polarisation? Which other mental glitches resulted in the disastrous Charge of the Light Brigade in the Crimean War and today haunt international trade negotiations and territorial disputes such as that between Israel and Palestine?

But we'll start by examining our evolution and see why, long before we cultivated wild plant species and tamed wild animals to create agriculture and civilisation, we first had to domesticate ourselves. How did humans develop to coexist harmoniously in larger and larger populations and cooperate successfully on shared ventures?

Chapter 1

Software for Civilisation

There is nothing to which nature seems so much to have inclined us, as to society.
—Michel de Montaigne, 'Of Friendship'

There are many advantages for animals to living in groups. It makes finding mates much easier; it allows for successful hunting in packs; and it offers safety in numbers and protection from predators. But compared to herds of wildebeest or schools of fish, there is a great deal more complexity in human societies. We have an incredible propensity to cooperate. The key to human success has been not just our adept tool use, made possible by the dexterity of our hands, but our willingness to offer a helping hand to one another, even if we're unrelated or unlikely ever to meet again. As Nichola Raihani puts it in her excellent book *The Social Instinct*, 'Cooperation is our species' superpower, the reason that humans managed not just to survive but to thrive in almost every habitat on Earth.' We teach one another skills and exchange information that we would never have worked out for ourselves in one lifetime. This process of cultural learning enables the spread of new capabilities not only throughout populations but cumulatively over generations.

In this chapter, we'll look at two major developments in human evolution that were key prerequisites for our ability to create complex, largely peaceable societies and work together in the huge enterprises that we call civilisations:* the reduction in reactive aggression and the development of the social software in our brains that enables unparalleled levels of cooperation.[1]

TAMING OURSELVES

It is simplistic to think of aggressive behaviour in terms of a single sliding scale from docile to violent. There are two forms of human aggression which are quite distinct from each other. Reactive aggression is a hot-headed response, an impulsive lashing out against an immediate threat. On the other hand, proactive aggression is driven less by impulse and emotion: it is calculated, premeditated action towards a specific goal. Throughout our development as a species, the expression of these two forms of aggression shifted in different directions – we evolved to be very moderate with the first, but highly proficient at the second. If we view aggression as a dualistic phenomenon, we can see that there is no contradiction in saying humans can be both brutal and benign.[2]

Our closest living relatives, chimpanzees and bonobos, live in mixed groups of males and females. These groups are fluid in their size and composition, splitting into smaller groups to forage different areas during the day, before reconvening to sleep at night. Over longer timescales, individuals move between different groups dispersed across the landscape; related chimpanzee males, for example, stick together but mate with females from neighbouring communities once they are old enough to breed.

This periodic division and reassembly of groups is known as fission-fusion social organisation. In such mixed groups of chimpanzees, outbreaks of aggression and violence are commonplace.

---

* For the purposes of clarity, what I mean by civilisation here is a complex social organisation, characterised by the centralised political and administrative state, a high degree of role specialisation, a stratified social structure, a distinctive cultural output and dense populations living in urban settlements.

Males harass females, and there is frequent antagonism and vicious competition between males over reproductive access to the females. Male in-fighting establishes a hierarchy, and the alpha male must use violence, or the threat of it, to maintain his position at the top. Male chimpanzees also form gangs to patrol the boundaries or their territory or invade that of neighbouring groups. They attack, and sometimes kill, males from other groups to expand their sphere and gain access to more resources or females. Bonobos are generally less violent than chimpanzees, but they also exhibit aggression both to other members of their group and to outsiders.[3]

While aggression is a way of life for chimpanzees, human evolution took a very different trajectory. The rates of physical aggression among other primates – even the more peaceful bonobos – are more than a hundred times higher than among humans.[4] Indeed, acts of reactive anger are remarkably rare within traditional hunter-gatherer societies today. These groups are also notably egalitarian, with no despotic alpha male or strong dominance hierarchy.

The key development in human evolution appears to have been the emergence of coalitions of males to keep in check or remove any would-be tyrant. There were two key drivers of this transition in our social structure: language and weapons. The ability to communicate effectively enabled individuals to conspire and plan a coordinated move against a tyrannical top dog, while reassuring one another of their shared intent and commitment. In short, language opens up the ability to plot the disposal of a despot. And when launching such an attack, the use of projectile weapons, such as a rock or spear, permitted a decisive move without any one individual exposing themselves to great physical risk.[5] Such coalitions tend to attack only when they have overwhelming numbers and are assured of victory. The same calculated mathematics of relative strength has been at the forefront of every general's mind throughout human history.[6] The first such planned killing of a despot would have happened hundreds of thousands of years before the assassination of the Roman dictator Julius Caesar in 44 BC.

The effectiveness with which individuals could join forces to safely challenge and dethrone aggressive despots levelled the playing field. An individual's influence within society became decoupled from their personal physical might, and instead came to rest on the strength of their social network and the reputation they had gained based on their generosity or supportiveness. Power shifted, from a dominant alpha male who acquired and then maintained his authoritarian position through brute force and the threat of violence against any challengers, to the wider group in a more equitable distribution. A new kind of political system had arisen and transformed the fabric of early human communities: strict hierarchy gave way to a more egalitarian structure. This reduction in reactive aggression and increased placidity of humans laid the foundations for the development of complex cooperation and cultural learning.[7]

This ability for coordinated alliances to keep violent despots in check with planned proactive aggression[8] created the selection pressure to reduce hot-headed reactive aggression. Unlike a chimpanzee in the prime of his strength, it no longer paid for humans to lash out at rivals in an attempt to rise to the top. Indeed, gaining a reputation for being violent only risked a coalition of opponents later rising against you. Collective punishment of reactive aggression resulted in its evolutionary suppression. We domesticated ourselves.*

* We can see a similar process in the domestication of wild animals. Compare any domesticated creature to its wild relatives – a dog to a wolf, a pig to a boar – and you notice, alongside an increased tolerance of humans, a marked reduction in reactionary aggression – the result of generations of selective breeding picking the traits for peaceful cohabitation.[9] The animals we domesticated exhibit more docile, placid behaviour compared to their wild relatives; they also tend to have a smaller amygdala, the part of the brain involved in the fear response and aggression.[10] (Domesticated animals also tend to exhibit a common cluster of physical traits including smaller muzzles and teeth, floppy ears and changes in pigmentation. On the whole, these were not targeted directly by selective breeding but are by-products of selection for less reactive aggressive behaviour.[11]) Interestingly, many of the genes that have been favoured in the domestication process of wild animals have also

As this shift in human social structure progressed, other, milder sanctions could be used to maintain balance within the group, without needing to resort to proactive violence. Anyone getting too big for their boots became the target of public ridicule, shaming or ostracism – we still find these patterns and rituals at work in hunter-gatherer societies today. But the threat of being attacked by a coalition of those who a dictator would attempt to dominate remained the ultimate deterrent. While the ability for a community to remove a despot does not guarantee an equitable and fair society, it is a prerequisite and goes a long way to levelling out a dominance hierarchy.

So, while hot-headed, reactive aggression was suppressed in the human evolutionary lineage, calculated, proactive aggression remained.[13] Surprise attacks from one settlement or village against another were motivated by the desire to remove competitors or acquire resources or mates. The more recent extension of such behaviour, emerging with the development of city states and civilisations, is all-out warfare. Indeed, war is the ultimate expression of proactive aggression, ordered by rulers, planned by strategists and commanded in the fray of battle by generals.

In normal life, lethal violence is socially prohibited; in war, on the other hand, the very objective is to kill a decisive number of the enemy. But humans generally have a deep-rooted aversion to discharging violence upon one another – a biologically encoded peacefulness borne of our evolution within egalitarian social organisations. While leaders may try to rouse their men with proclamations about the honour and glory to be had on the battlefield – fighting for God, king or fatherland – many soldiers throughout history, many of them farmers mustered from their fields, have found the thought of killing another person utterly abhorrent. The social traits and inclinations that enabled humanity to live harmoniously in complex societies and develop civilisation

---

been positively selected in the human lineage since our evolutionary split from the Neanderthal lineage over 500,000 years ago, reflecting common genetic changes relating to brain function and behaviour.[12]

must be overcome to prepare us for war. In order to induce troops to kill, military training is often directed towards increasing aggression, and propaganda aims to dehumanise the enemy.[14]

CIVILISATION AND THE RE-EMERGENCE OF THE DESPOT

This largely egalitarian social structure is believed to have held sway for the great majority of our history as anatomically modern *Homo sapiens*. But the desire for personal power and dominance never disappeared. Indeed, the conditions created by the introduction of agriculture and the arrival of the earliest civilisations led to the re-emergence of despotic supreme rulers.

In a hunter-gatherer society, the fresh meat from a successful kill, or foraged perishable plant products such as fruit, must be consumed immediately before they go off, so it makes sense to share them among the group. In any case, the group is constantly on the move and does not have the capacity to accumulate a stockpile of resources.

With the development of agriculture, humans began to live in permanent settlements alongside their fields or pastures. Farmers were no longer limited to the possessions they could personally carry. What's more, the glut of food produced at harvest time, and the need to store the surplus in granaries, created commodities that could be hoarded. Thus was born the concept of wealth. Agricultural surplus enabled denser and denser concentrations of humanity, the emergence of cities and greater levels of social organisation, leading to more complex states and the development of civilisation.

While there is evidence that some hunter-gatherer populations were not perfectly egalitarian and did exhibit degrees of sedentism, social stratification and specialisation of roles within the community,[15] it is clear that these characteristics all became much more widespread and pronounced with the advent of agriculture.

   Individuals who assumed a position of leadership, perhaps through their skill in rallying peers to work together in successful shared enterprises like constructing and maintaining irrigation systems, were able to exercise authority over such vital infrastructure and accumulate resources for themselves. Those exerting control over the distribution of valuable caches of food and other assets could withhold resources to exercise leverage or use them to buy allegiance to quash leadership challenges or uprisings. And through the transmission of material riches and social rank from one generation to the next by familial inheritance (something we'll come back to in the next chapter), initially small differences in resource wealth – and the influence and stature it affords – came to be amplified. Rulers were able to consolidate their position; privilege and power became increasingly concentrated in elite classes and the social structure ever more stratified. In an agricultural world dependent on established infrastructure and city life, people were less able to simply move away and had little choice but to put up with increasingly autocratic rulers.[16]

   This disparity of power was only exacerbated by the innovation of the first metalworking processes and the production of bronze weapons, armour and shields. In the ancestral condition, the general availability of potential weapons – any heavy stone or pointed branch could be wielded against an enemy – fostered egalitarianism. But when superior weapons and armour are difficult to manufacture, or the raw materials rare and expensive, the effect is to bolster the dominance of the despot. Only the top dog controlling the wealth can afford to buy the loyalty of fit, strong men and equip them with cutting-edge arms technology. It becomes a great deal harder for an ad hoc coalition of individuals to remove a tyrant. Indeed, a state is often defined as a coherent polity that is able to operate a monopoly on violence within its boundaries – with the sovereign ruler controlling where and when that violence is directed.[17]

COOPERATION AND ALTRUISM

We have not only modified our patterns of aggression to live peaceably together in large groups, but become prodigiously cooperative and uniquely altruistic. It's important to distinguish between the two: altruism delivers a benefit to the recipient at the expense of the donor; whereas cooperation benefits both parties. Cooperation is widespread in the animal kingdom. Hyenas working in packs to bring down an antelope far larger than themselves collectively achieve a goal that no one individual could on their own. But the sheer extent of cooperation exhibited by humans overshadows anything like that of any other species on the planet. Civilisation is itself the ultimate expression of cooperation – of large groups of people contributing to the same shared venture.

Much of the assistance that humans give one another is altruistic. This means that one individual helps another at a cost to themselves – in terms of food, energy, time or other valuable resources – seemingly with no immediate personal benefit. At first consideration, such acts appear to be difficult to explain within the context of evolution. If every individual in a population is in competition with others to survive and reproduce, what can be gained by helping another, especially at a cost to oneself?

Natural selection is often thought about in terms of an individual's ability to survive in a particular environment, compete against members of their own species as well as others and succeed in finding food and mates. Those with advantageous traits prevail and reproduce, so in the next generation more individuals carry the particular genes that produce those traits, and over time a species adapts to become better suited to its environment. The real success of an individual isn't just the number of progeny they are able to produce but the number of progeny that themselves survive to go on to reproduce. It's taking the long view: fitness is about maximising the number of *grandchildren* you have.[18]

But there's another key insight here. Selection favours not only traits that advantage your own direct descent – your number of grandchildren – but also those that contribute to the reproductive success of relatives. A particular gene propagates not only if a given individual who carries it gains an advantage, but also if related individuals – who are likely to be carrying copies of the gene – survive and reproduce. This is the concept of inclusive fitness.

By this rationale, an individual can help copies of their own genes survive and spread by assisting their relatives, in proportion to how closely related they are. More specifically, an individual's genes will prosper if the cost incurred by the individual in helping a relative divided by the benefit received by that family member is less than their degree of genetic relatedness. This is known as Hamilton's rule, after evolutionary biologist W. D. Hamilton, who expressed it in a mathematical formula. But it's best understood with an example. You are 50 per cent genetically related to a full sibling – which is to say, there's a 50:50 chance that any randomly picked gene of yours is identical in your brother or sister through descent – and provided that any action you take to help gives them at least twice as much benefit as it costs you, then it will lead to an overall advantage for your shared genes. This key realisation led evolutionary biologist J.B.S. Haldane to quip once to friends in a London pub that he would jump into a river and risk his life to save two brothers, but not one, or to save eight cousins, but not seven.[19] By helping your family members, particularly if they are close relatives, you are indirectly serving your own genes. This evolutionary strategy of assisting the survival and reproduction of relatives, even at a cost to oneself, is known as kin selection.

Apparently altruistic behaviour directed towards relatives is still self-serving, therefore, in that it helps to propagate the genes that you share. In small, close-knit communities, with little coming and going of individuals from other groups, the people surrounding you are likely to be related and so it pays to be generally helpful towards other individuals in your own group.

Kin selection is everywhere in the animal world: many species have been shown to preferentially help their immediate family, or those in their group who are odds-on likely to be related and so share many genes. What's more, many animals, including humans, appear to possess an encoded appreciation of Hamilton's rule: they not only behave more altruistically to kin compared to non-kin, but are also more altruistic towards closer kin than more distant relatives.[20] Within human populations, kin selection expresses itself in everything from charging towards a predator to protect family, going hungry to feed siblings or helping to raise a sister's youngsters (or diving into a river to save a particularly unfortunate group of eight cousins).*

Kin selection provides a neat explanation for most of the altruism we find in nature. But it cannot account for acts of generosity towards non-relatives. How can a behaviour be evolutionarily advantageous if it comes at a cost to yourself, but you cannot count on the beneficiary sharing any of your genes? The fact that, compared to other animals, humans are freakishly kind to non-relatives demands another explanation.

RECIPROCAL ALTRUISM

The theory that is generally believed to explain how non-related individuals may benefit from helping each other is known as reciprocal altruism. The idea is that if one individual helps another, even if paying a cost in doing so, the favour is returned at a later time. In this way, cooperation can evolve as a series of mutually altruistic acts.[22]

---

* One form of kin selection has become known as nepotism, originally describing favouritism granted to relatives. The word itself derives from the Italian for 'nephew'. Catholic bishops and popes, in making influential appointments, would often favour their relatives, often nephews. As the next pope is elected by the cardinals, this enabled them, despite having taken vows of celibacy, to attempt to continue their own dynasty.[21]

Such reciprocal altruism isn't nearly as common among non-human animals as kin altruism, but there are examples in a few species that, like humans, have an ecological necessity for social interactions.[23] Evidence for reciprocal exchange can be found among other primates, including baboons and chimps, as well as among rats and mice, some birds and even fish.[24] One of the best-studied cases is that of vampire bats. These bats feed on the blood of large wild mammals, as well as of our domesticated livestock. But finding a meal can be difficult, and because of their high metabolism, these animals need to feed every day or two. Vampire bats live in large groups, and if one individual has successfully fed it will often regurgitate blood to share with a less fortunate colony-mate. A bat that altruistically shares blood one night is likely to have the favour returned another day when the tables have turned.[25]

There's a simple economic principle lying at the core of why reciprocal altruism works so well. Those who successfully gathered food often have acquired more than they need to survive. The surplus becomes less valuable, making only a marginal difference to their prospects. But for an individual that does not yet have enough to eat, that extra unit of food is still very valuable – it could mean the difference between life and death. So a benefactor can donate some of their surplus to someone in need at minimal cost to themselves but huge benefit to the recipient. In the case of the vampire bats, one feast on an animal supplies more than enough sustenance and so an individual who successfully foraged has food to bestow on another, less fortunate bat who may otherwise have starved to death. Later, when fortunes have shifted and the original recipient has a surplus, they can return the favour, again with the greatest possible utility being extracted. Thus reciprocal altruism is a form of asset exchange, and each donor can receive a profitable return on their investment.

By engaging in this practice, both parties have extracted the maximum value from a surfeit they possessed at different times. For this reason, the behaviour is often also called delayed altruism. Competition is said to be a zero-sum game: for one individual to gain, another must lose. But cooperation is non-zero-sum: both sides can profit from the arrangement and often substantially so. This dynamic is utilised by both vampire bats and early humans sharing food and other resources or performing a service for one another. As Raihani points out, 'Reciprocity is so fundamental for driving cooperation that it has become enshrined into well-known proverbs. *Quid pro quo*. You scratch my back, I'll scratch yours. Do as you would be done by. One good deed deserves another. These maxims exist in other languages too. In Italian, *una mano lava l'altra* translates into the particularly lovely "one hand washes the other", a phrase that also exists in German (*ein Hand wäscht die andere*). In Spanish, *hoy por ti, mañana por mi means*, roughly, "today for you, tomorrow for me".'[26]

The problem with altruistically providing resources or services willy-nilly, helping others when you can't be certain that they will reciprocate in the future, is the risk of being played for a sucker. Cheaters can take advantage of your indiscriminate generosity, and you end up paying all the costs of helping but receive few benefits back. For the system to work, freeloaders must be held in check: those who don't reciprocate need to be punished by being refused help next time so as to incentivise mutually cooperative behaviour. If the recipient refuses to repay the kindness when fortune swings in their direction, the original altruist needs to remember and desist from helping them again in the future: once burned, twice shy. This tit-for-tat behavioural strategy is also found in some animals: ravens, for example, have been found to refuse to help other individuals who cheated in the past.[27]

FRIENDSHIP AND THE BANKER'S PARADOX

Keeping a mental ledger on which individuals did or did not reciprocate favours carries its own cognitive burden, however, and human evolution has devised a solution to this. After repeated rounds of reciprocation with the same individual we begin to relax our monitoring of the exchanges. In other words, we come to trust one another, and the relationship develops into a deeper bond: friendship. A friend serves as a trusted collaborator and ally in other social interactions, and we suspend the mental accountancy on keeping track of their behaviour and no longer explicitly expect or demand any particular favour to be repaid. The bond is its own assurance of reciprocity and an investment in the future.[28] We know that friendships sour, of course, but only after a long history of one partner taking more than they give back.

The bond of friendship is biologically mediated through oxytocin, the hormone that serves in all mammals to drive maternal care of their young, and in humans sustains the pair-bond between sexual partners long enough to successfully raise children together (which we'll come back to in Chapter 2). Friendship among humans is an extension of this close-knit relationship between parents and their offspring: we also forge a tight bond to those with whom we regularly reciprocate. It is this neurochemical bond that makes the pain of betrayal by a close friend so much more intense than the vexation of being cheated by a stranger.

In particular, the bond of friendship may solve a problem known as the banker's paradox. When you are facing financial ruin and most need a loan, the bank is unlikely to grant you one as you represent a terrible credit risk. On the other hand, when things are going well the bank is only too happy to offer you funds. This same dynamic would also have posed a deep problem for reciprocal altruism in the world of our ancestors. Individuals may be least likely to receive help when they most need it, because they are least able to reciprocate. Why would a non-relative come to your aid, with a greatly reduced chance of being paid back the favour? The evolution of friendship provides a solution to the quandary. The oxytocin-mediated bond between friends makes them irreplaceable to each other. So if a friend falls seriously ill, rather than callously abandoning them to find someone else with whom to engage in reciprocal altruism, you have an emotional stake in their well-being that compels you to help them pull through. A friend in need is a friend indeed. In this way, friendship may have developed in human evolution as a form of insurance against desperate times.[29]

There are a few known examples of reciprocal altruism in the animal world – such as among the vampire bats – but the practice is exceptionally common among humans. It accounts for a great deal of the generosity and cooperation seen in our interactions, especially within small, tight-knit societies where individuals have a high probability of encountering one another again so altruistic deeds can be repaid. But one extraordinary feature of human behaviour, compared to all other animals, is our propensity for helping each other even when there can be no expectation of regular interactions. This is the kindness of strangers. Humans often ungrudgingly offer assistance even to those they have never met before and can have no great expectation of ever encountering again. How can such one-off acts of kindness be explained? Kin selection and reciprocal altruism cannot account for this behaviour; there must have been other things at work in the development of our species.

One possible explanation is an evolutionary mismatch. The ancestral human condition was life in small bands with most individuals related to one another. Under these circumstances, kin selection and reciprocal altruism can comfortably explain generous acts between tribemates: you're either directly helping copies of your own genes or repeatedly interacting with the same individuals for a favour to be returned. But this simple evolutionary strategy would have no longer worked so well as humans began living in larger, more complex societies, particularly when ever-greater populations settled in urban environments, dominated by fleeting interactions with strangers with no familial connection. On my morning walk into work, I pass more strangers on the street than my hunter-gatherer ancestors probably encountered in a lifetime. Yet in general we continue to cooperate with those around us, even though there is no longer any genetic self-interest.

Our minds evolved to drive behaviour that was adaptive in our ancestral conditions, in small, kin-based communities on the African savannah, and this cognitive operating system has not had a software update as the social environment has rapidly transformed. Thus our altruistic dispositions are not calibrated to our evolutionarily novel world. This produces the apparently maladaptive behaviour of helping strangers when the favour will never be returned by them. [30]

But there's a better explanation for why humanity is so prolifically cooperative without an expectation of direct reciprocation – and it actively explains this apparently paradoxical behaviour rather than just seeing it as an evolutionary programming hangover.

INDIRECT RECIPROCITY

The notion of indirect reciprocity holds that rather than returning a favour to the same altruist, the benefactor pays it forward to others. A helps B, who then helps C, who then helps D and so on. The favour is transferred around the community, until sooner or later it returns to A. What goes around comes around. And there's an additional level: another individual who witnessed A's initial act of kindness towards B helps A themselves in order to build a relationship with somebody they know to be generous: Z helps A. The same two individuals don't need to encounter each other again, as is required for direct reciprocity, but benefit from the altruistic behaviour of the group as a whole. Helpful people are themselves more likely to receive help, whereas freeloaders who refuse to help others are punished or excluded.[31] Such indirect reciprocity is a uniquely sophisticated form of human cooperation,[32] and for the system to work, it needs two crucial functionalities not possessed by other animals.

Firstly, not only must there be witnesses observing interactions, and whether either party acts generously or selfishly, but that valuable information on the behaviour of individuals must be shared in a common pool of information for the entire group. In other words, members of a community must gossip about one another. If an individual becomes known for unreliability, for selfishly accepting benefits and not helping others, members of the community will simply withhold aid the next time the swindler is in need. It's not quite true that 'cheats never prosper' – they can often get away with it in the short term, especially in larger, more anonymous communities – but they are caught out sooner or later and their reputation is stained. Gossip, therefore, is a key prerequisite for ensuring indirect reciprocity doesn't become overburdened by freeloaders, and its omnipresence in human cultures spans from the campfire to the water cooler. Indeed, gossip and chit-chat came to replace other relationship-forging activities in primates, such as grooming.

This prolific sharing of information throughout the community on each member's behaviour – like a social internet mediated by chit-chat – creates a reputational system for determining every individual's suitability for cooperation attempts. An individual who acts generously to others develops a good reputation; an unreliable freeloader gains a bad reputation, and others know to avoid them in future interactions. Natural selection favours an individual who acts kindly because others are inclined to help them later, and so evolution has crafted human psychology to makes us care deeply about our reputation, while gossip keeps us playing fair.

The first rule of life in a gossiping society is to be careful what you do – or, more importantly, to be careful about what others will think about what you do.[33] Human society thus became a crowd of minds simulating other minds – inferring the motivations and attitudes of others and how they are likely to perceive your actions so that you can better manage your reputation. Our conscience is an expression of this – it's our inner voice that warns us someone might be watching and makes us consider how they would likely perceive our action so that we can avoid social punishment.[34]

The second crucial facilitator of indirect reciprocity is the punishment of cheats. In the repeated one-on-one interactions of direct reciprocation we looked at earlier, an individual remembers if another person cheated them previously and so can refuse help next time. Chimpanzees are also known to take revenge for acts that personally disadvantaged them.[35] However, a behaviour unique to humans is a party who wasn't directly involved in an exploitative interaction punishing the cheat for no material gain to themselves – something that is known as third-party, or altruistic, punishment.[36]

Altruistic punishment behaviour in humans can be explored with simple economic games. The kind I'll discuss here involves a group of players contributing to a collaborative outcome that is advantageous to all – what is known as a public good. Such cooperative endeavours are ubiquitous in human societies, from hunting a large kill to digging and maintaining a channel system to irrigate farmers' fields to constructing a municipal building. The history of civilisation is the history of people contributing to public goods, and as civilisation has advanced, the number and complexity of public goods has increased accordingly.[37] Cities and states provide services such as decent roads, a clean water supply, emergency services, public education, healthcare, law and order and national defence. The outcome can be enjoyed by the entire community, but only those who participated bore the costs.

Public goods are vulnerable to being undermined by slackers who may get away with putting little to nothing towards the shared venture but still reap the rewards. The public good game is often set with each player having a pot of money, and in each round of the game they can choose how much to contribute to a communal pool. At the end of the round, players pocket what they had kept behind in their personal pot, and the shared pool is multiplied by some factor (between one and the number of players) and distributed evenly among everyone. The best possible outcome for the group as a whole is for all players to contribute their entire pot so that everyone maximises the multipliable funds and thus their individual returns. But a free-riding player can profitably cheat on the cooperative effort and pitch in nothing – keeping not only their own entire pot but also the dividend of everyone else's generosity.

What tends to happen is that most participants contribute about half of their personal pot to the shared pool – a reasonable, cautious approach. However, as the players realise that some of their number are putting very little into the shared pool, or even nothing at all, everyone's contributions decrease round on round towards zero.[38] The cooperative venture collapses under the self-serving actions of freeloaders.

But there's a simple modification to the rules of the game that can enforce cooperation and rescue the shared venture to everyone's benefit. The addition of a sanctioning system allows players to spend some of their own game money to reduce the income of those they felt had cheated – for example, they can pay £1 to reduce a cheat's take-home by £3. The inclusion of such altruistic punishment radically changes the dynamics of the game. Now the individual contributions towards the common good tend to rise – sometimes to over 70 per cent of each individual's pot – and remain at that level round on round. It seems that people are willing to incur a personal cost to punish cheaters, and this altruistic punishment is very effective at both deterring free-riders and encouraging greater cooperation among the group as a whole. And so in real life too, inveterate cheats whose selfish or antisocial actions sabotage the community risk punishments including the denial of benefits, social exclusion or ostracism – or may even become the target of proactive violence.

The key motivation driving altruistic punishment seems to be innate and emotional – players report feeling indignation or anger towards the free-riders and an impulsive desire to punish them.[39] Studies have found that the righteous punishment of cheats triggers the same surge of the neurochemical dopamine in the reward centres of the brain as crucial biological functions such as sating hunger or thirst, having sex or providing parental care. (We'll come back to the dopamine system in Chapter 6.) It is this rush of dopaminergic pleasure that drives us to shoulder the costs of delivering deserved punishment to others.[40] In humans, it would seem, enforcing cooperation and prosocial behaviour has its own innate reward.[41] On top of our primal urges directed towards survival and reproduction, therefore, are layered more recent neurological compulsions for our extraordinary prosocial behaviour. Humans appear to be hardwired with cooperation as the default state, and they demand fair play in interactions. Cooperation is in our coding and altruistic punishment is the glue that holds societies together.[42]

For indirect altruism to function in a society, the burden of administering such punishment must itself be shared among all members of the group. So we take things one step further still and also punish those who shirk the responsibility of sanctioning transgressors (and who are themselves therefore effectively free-riding on the reputational system).[43] Humans are not just altruistic; we are vigilant enforcers of altruism.[44] Altruistic punishment is a necessary condition to prevent cooperation being undermined by free-riders and explains how indirect reciprocation evolved in the first place.[45]

Indirect altruism demands sophisticated cognitive abilities. Not only does the reputational system require language and the sharing of information through gossip on which individuals have proven themselves reliable or unreliable cooperators, but each individual in the group needs to keep a mental ledger of how reputable is each of their peers in the community.

Humans have a finite capacity for keeping track of all this social data, however. The suggested cognitive limit for the number of long-term social relationships that an individual is able to maintain (which thus also limits average group size), known as Dunbar's number, after the British anthropologist Robin Dunbar who first proposed it,[46] is usually estimated to be around 150.* We have a few very close, trusted friends and relatives, and outside this core group, we surround ourselves with

* Some estimates place the maximum network size as twice that[47] or more.[48]

progressively larger concentric circles of individuals we know to lessening degrees, down to acquaintances that we interact with only rarely. The numbers within each of these circles seem to remain fairly constant, so that if we make a new friend, we tend to lose contact with an old one we haven't seen in a while. This layered structure of our social networks has been identified in different modern societies[49] and is evident in who we call or text on our phones and how often,[50] how we interact on social media[51] and even how we play online games.[52] Although our modes of communication have changed with new technology, our Palaeolithic brains have not.

In larger societies, it is these personal social networks, existing as fuzzy-edged, over-lapping domains within the total population, that foster clusters of altruism and cooperation within a world of strangers.

DETECTING CHEATS

As well as experiencing an innate impulse to punish those who break the rules on reciprocating altruism or cooperation, humans are adept at detecting instances when such infractions occur. The importance of catching cheats is so central to protecting our cooperative social life that we have developed a particularly sensitive ability to identify when rules have been broken.

We don't normally fare very well at tasks involving the application of logical rules. A classic puzzle used to explore this is known as the Wason selection task.[53]

Imagine you are sat at a table with four cards laid out in front of you, as shown below. Each has a number printed on one side and is coloured either black or white on the other side. You are told that if a card bears an even number then its opposite face is white.

Which card or cards must you turn over in order to determine if the rule is true or not? This demands what is known as

falsification-based logic – to find the truth, you must try to prove that a conditional rule (if p then q) has been broken.

The correct answer is that you need to check both the card showing the '4' and the one with a visible black face. Turning over either the card with a '7' or the card with a white face is in fact irrelevant to the puzzle – even if they do bear the 'wrong' reverse face, it doesn't falsify the rule, since the rule doesn't say that cards with odd numbers can't also have a white back, for example.

If you got this wrong, don't worry: you're firmly within the majority. Only around 10 to 25 per cent of people presented with this problem get it right.[54] Most recognise that finding confirmatory evidence of q in the presence of p – i.e. that the reverse face of the '4' card is indeed coloured white – is important; but they don't turn over the black card to look for falsifications of the rule – a p (even number) without a q (white face).

But, amazingly, if the same logical test is couched in terms of detecting cheats in a social exchange – if you take the benefit, p, then you must pay the cost, q – people perform a great deal better.[55] Let's say that a local farm has a table by the side of the road with pumpkins priced at £1 each; there's an honesty box that you can drop your coins into. The cards for this puzzle are shown below: one side displays what an individual paid, the other whether they helped themselves to a pumpkin or not. Which would you turn over to check whether the individual represented by each card had cheated the honesty system?

| Paid £1 | Paid nothing | Took a pumpkin | Did not take a pumpkin |

In this case, the puzzle seems pretty straightforward, facile even. But it represents exactly the same process of falsification-based logic operating on a conditional rule that we encountered in the earlier example. I suspect that you, like me, would immediately reach to check the reverse side of the 'Paid nothing' card to discover whether the individual dishonestly took a pumpkin anyway, but also the 'Took a pumpkin' card to see if they had paid the required pound. We seem to intuitively know how to verify the rule of 'if you take the benefit (p) then you pay the cost (q)'. When the Wason selection task is framed as a social exchange, 75 per cent of people give the correct answers – a rate over three times higher than for the abstract version with numbers and colours.[56]

Perhaps it's not a complete surprise that people perform better at this task when it is framed in terms of a real-life, familiar scenario. But even when the Wason test uses everyday concepts – though importantly not related to fair play – people still perform badly. Yet children as young as three, when presented with the challenge in an age-appropriate pictorial form, correctly identify 'naughty' people violating social exchange rules by taking a benefit without meeting the requirement.[57]

Some psychologists and anthropologists argue that this demonstrates humans have an innate, specialised 'cheater detection' module in our brains which has been tailored by evolution to detect violations of fair, cooperative behaviour.[58] The claim is controversial, however, and difficult to prove,[59] but what is clear is that humans are especially good at catching free-riders, and this has been a critical facility in enabling widespread cooperation in large societies.

FROM SOCIETIES TO CIVILISATIONS

Evolution has equipped us with a set of internal drives that moti-vate beneficial behaviour: a mounting feeling of hunger compels us to eat; sexual desire and the prospect of orgasm incentivise us to reproduce. Evolution has also moulded our tendencies to pro-mote behaviours that brought us advantages in living together in groups. These biologically encoded responses – which we perceive as emotions – include affection for family and friends, sympathy for those who are suffering, indignation and anger towards cheats, and the warm glow of satisfaction from performing an altruistic act or delivering righteous punishment. Other emotions promot-ing social cooperation are self-directed. Distressing feelings of guilt and remorse suggest we have acted unvirtuously, and express-ing these feelings to the community can serve to mitigate the social punishment and help pave the way to mending relationships and being forgiven.[60] These emotions are deeply embedded in our cog-nitive software – they were probably already in place in some form before our divergence from the chimpanzee lineage – and they promote altruism, reciprocity, cooperation and fairness as the building blocks of human morality.[61]

Morality provides a critical framework for living harmoni-ously in social groups. Most people around the world would agree that helping others, keeping promises and remaining faithful to your spouse are moral behaviours, whereas murder, rape and deceit are immoral.[62] In broad terms, immoral behav-iour can be defined as advancing one's self-interest at the expense of other individuals, including committing an act with-out their consent or removing their agency to make their own free choices. Most of what we consider moral behaviour is that which is unselfish and conforms to fairness and cooperation in social exchanges so as not to destabilise society.[63] The core of moral behaviour is the Golden Rule, 'Do as you would be done by', which requires an individual to consider their actions from another person's perspective.

These innate human impulses promoting altruistic and cooperative behaviour, and the sense of morality that grew from them, maintain prosocial behaviour in small communities. But as group sizes grow, monitoring cooperation becomes more difficult. Direct reciprocity becomes less effective because cheats can keep finding new marks and not face punishment from a repeated interaction. Indirect reciprocity also breaks down because within bigger groups information can be slow to diffuse and it's harder to keep track of people's behaviour. Relative anonymity in large populations allows cheats to keep ahead of their own bad reputation.

A society can only grow so large before the innate mechanisms promoting prosocial behaviour in humans become insufficient and cooperative endeavours risk collapsing altogether under the burden of cheats and freeloaders.[64] In order to enable the large populations of cities and civilisations to exist more peaceably, cultural constructs have emerged to supplement the biologically evolved drivers for cooperation.

Religion, and the belief in gods, is one such cultural innovation that has had a huge influence since the origin of civilisation. Gods have been invoked as many things in human cultures: creators of the Earth and the cosmos; causes of natural phenomena; interventionists in events both propitious and disastrous; and determinants of individual fate and fortune. But it is their role as all-seeing observers and all-powerful disciplinarians (as well as mediators of forgiveness) that serves as a powerful incentive for prosocial behaviour in large communities where transgressions may otherwise go undetected. An omnipresent, omniscient and omnipotent god serves as an ultimate third-party punisher of immoral acts. While belief in supernatural beings and religion is not necessarily a prerequisite for forming large and complex societies and civilisations, it certainly helps.*

* I happen to be an atheist myself, but as a scientist and historian I can acknowledge the beneficial functions of religion as a social construct without having any personal spiritual beliefs. Indeed, belief can have positive (or

The invention of writing coincided with the emergence of the first city states. Writing is a technology that seeks to overcome the limitations of human memory and oral communication for transmitting knowledge throughout society and down the generations. The first writing systems used in Mesopotamia served to record details of civic administration and trade around the fourth millennium BC; in Egypt and Mesoamerica, it was used to create calendars and chronicle the history of political and environmental events. Crucially, writing was co-opted by the state for establishing codified laws. The earliest known written codes are from Mesopotamia. Surviving fragments of the codes of Ur, a Sumerian city-state, from the third millennium BC provide a list of punishments for different crimes in the form of 'If ... then' conditional rules. The punishment for agricultural crimes was fines in barley; fines of silver were prescribed for bodily harm; robbery, rape and murder, however, were capital offences. The Code of Hammurabi, King of Babylon, from around 1750 BC is particularly well preserved and consists of over 4,000 lines of cuneiform text inscribed into a stone stele that was displayed in public. It covers areas of law including family, property, trade, assault and slaves, again written as 'If ... then' conditional rules. Provisions include: 'If a man should blind the eye of another, they shall blind his eye,' and 'If a man breaks into a house, they shall kill him in front of that very breach.'[66]

Most actions which have been proscribed by systems of law around the world and throughout history are those which our evolutionarily forged, collective sense of morality already denounces. These are crimes against other people and their belongings, such as assault, murder, rape, and theft or damage of another's property. They prohibit slander – that is, falsely harming another's reputation – as well as reckless or negligent

---

negative) effects independent of whether a god exists or not. Yet although faith often encourages socially beneficial behaviour, religion is not the source of morality: a sense of morality appears to be innate in humans.[65]

behaviour. The Code of Hammurabi, for example, contains the somewhat draconian law that if a builder constructs a house that is not firm, and it collapses and kills the inhabitants, the builder shall himself be put to death. Then there are more recent crimes of non-compliance, such as failure to pay taxes – and thus cheating in the public good that is society itself. Non-compliance crimes are sometimes referred to as victimless, in that no specific injured person can be identified, but they potentially harm everyone in society.

Fundamentally, therefore, legal systems alter the social environment in order to modify human behaviour. They further incentivise prosocial attitudes in large societies of strangers by making it more likely that cheats are caught and penalised. Law is a tool to move humans to behave in ways they may not have if they believed they would get away with it.[67] Courts emerged to establish culpability and issue punishments such as fines, imprisonment or, for the worst social transgressions, capital sentences. More recently, police forces formed to detect wrongdoing and enforce the laws. Cheater detection and punishment still requires a collective cost to be incurred, but today every member of society contributes to the public good by paying taxes, and these are used to pay the salaries of police officers, court officials and prison guards.

In situations when the rule of law was not reliably enforced, the ancient system of trust based on an individual's reputation was extended to institutionalised reputational systems. In *The Social Instinct*, Raihani gives the following example. 'Traders in the eleventh century faced a dilemma when it came to selling their goods overseas: they could personally travel with their wares and sell them on the foreign market themselves, or they could entrust the task to a foreign agent, who would sell the goods on their behalf. The latter was a more efficient option – but carried the attendant problem of trust: how could a trader be sure that the foreign agent wouldn't just take the goods and clear off with them? The solution came in the form of merchant guilds, such as the Maghribi traders – a club that only admitted the most trustworthy members of society. By choosing to do business with a member of the Maghribi guild, a trader could be sure that his partner was committed to doing honest business. A Maghribi trader would face the much larger cost of being excluded from the guild if he didn't toe the line. People intuitively trust the drivers of London's famous black cabs for the same reason.'[68] The threat of being stripped of their prestigious licence far outweighs the short-term gain of swindling a customer out of a few quid. Today, the use of reputational systems for facilitating transactions between strangers has gone digital, with the burgeoning online peer-to-peer economy being supported by the practice of providing public reviews and star-ratings on online marketplaces and service-providing platforms (such as AirBnB, Uber, Lyft, TaskRabbit).[69]

The key question explored in this chapter relates to the intrinsic aspects of human nature: are we innately peaceful or violent? Two famous philosophers, Thomas Hobbes and Jean-Jacques Rousseau, held opposing views on the issue. Writing in the mid-seventeenth and mid-eighteenth centuries respectively, neither had the archaeological or anthropological evidence available today regarding how our ancestors lived in the distant past, so they postulated on the lifestyle of ancient humans before the arrival of civilisation. Hobbes believed that our natural condition was a precarious existence on the brink of survival, with humans in a perpetual struggle against each other and in constant danger of a violent death, before the emergence of a powerful state – Hobbes called it Leviathan – brought this barbarity under control. Rousseau, in contrast, argued that violence is not an intrinsic aspect of human behaviour and that our ancestors lived in idyllic harmony with each other and their bountiful environment, with no need for conflict. Humans, he reckoned, are naturally good; it is large, organised societies that have corrupted us.

As is so often the case, the truth appears to lie somewhere in the middle of this dichotomy. As we've seen, over our evolutionary history we domesticated ourselves and developed to repress reactionary aggression. Coalitions of individuals used the threat or, if necessary, the actual delivery of proactive violence to oust despots. Hunter-gatherers lived in largely egalitarian communities, although violent conflict frequently occurred between groups. The emergence of settled agriculture meant that property could be hoarded, which soon resulted in increasing disparities of wealth and power and the appearance of severely hierarchical social structures, which allowed the strong to exploit the weak. But the top-down control and monopoly of violence exercised by emerging states also fostered more peaceable existence in larger populations, and although states waged war with one another, the coalescence into larger polities and empires maintained greater order and reduced conflict within them.[70]

The crucial engine that enabled humans to work well together in ever-larger, more complex societies, eventually leading to the establishment of entire civilisations, was a progression of increasingly sophisticated systems for fostering altruistic behaviour and cooperation between individuals, as well as safeguarding against freeloaders. Kin selection functions perfectly within small groups of interrelated families. Direct reciprocity broadens the scope to also support cooperation between non-kin. And indirect reciprocity enables cooperation in even larger groups, facilitated by reputational systems, third-party punishers, and trust established among social networks within the wider population. All this is enabled by the social software that has evolved within our brains. But it is not enough to support larger societies, so civilisation must be held together by cultural inventions laid on top of our intrinsic sociality and desire for cooperation – such as religion, formalised codes of law, state-administered monitoring and punishment of wrongdoers, as well as institutionalised reputational systems such as merchant guilds.

While cooperation among peers had enabled coalitions to remove tyrants earlier in our history, the social structure within civilisations became increasingly stratified. Individuals came to amass material possessions and exert control over the vital agricultural infrastructure and the distribution of resources, and so were able to buy allegiance and quash uprisings. Initially small differences in dominance became amplified and fixed. Other cultural developments, such as the forging of metal weaponry and armour, further concentrated the use of force, and states were able to establish a monopoly on violence within their boundaries and keep their subjects in check. Those at the top of the social pyramid could consolidate their position, and leaders became rulers became despots. And with the familial inheritance of material wealth and social status, the configuration of privilege and power was passed down the generations.

It is the influences of the family in human history to which we will now turn our attention.

Chapter 2

Family

The family ... must be regarded as the natural, primary cell of human society.

—Pope John XXIII

Our closest living relatives today, chimpanzees and bonobos, go about their lives in a manner that is probably very similar to our common ancestor. Living in a forest environment, they spend their time swinging from branches and moving across the ground largely on all fours. As our human lineage diverged, we became increasingly adept at walking upright on our legs – we developed bipedalism. As the forested biomes in our evolutionary cradle of East Africa dried out and were replaced by grasslands, this allowed our ancestors to successfully inhabit the savannah and ultimately disperse around the world.

Alongside the development of bipedalism, a second major change was playing out: our ancestors were getting smarter. Over time, hominin species developed larger and larger brains – as can be seen from the increasing volume of the cranial cavity in fossilised skulls – and therefore were capable of progressively greater levels of intelligence, leading up to the exquisite aptitude that our modern species, *Homo sapiens*, has for language, cooperation, problem-solving and tool-use.

There's a problem, though. The two developments of becoming bipedal and brainy aren't readily compatible with each other. Mammals give birth by pushing the foetus out of the womb through a hole in the middle of the pelvis. Yet the adaptations to our skeletons and pelvis that enabled us to walk upright are in conflict with the need for a wider birth canal to accommodate the passage of bigger skulls. The human species today effectively sits right on the cusp of these two mutually exclusive design principles.[1]

The mechanics of birth, therefore, threatened to limit how brainy our ancestors could become. The solution hit upon by natural selection was to extend our developmental process long after we emerged from the womb. Compared to other mammals, including the other great apes, human babies are born in a remarkably undeveloped and vulnerable state. While a zebra can effectively get up and walk alongside its mother mere minutes after birth, humans take many years to be able to walk, feed and fend for themselves. All newborn mammals are fed milk by their mothers – indeed, the name for this group of animals comes from the Latin word *mamma*, meaning 'breast' – but the dependence of human babies goes much further than this. It is only after we pass through the hoop of birth that our brains are free to grow, and it is during these tender, formative first years that we learn how to coordinate ourselves and walk, talk and use the intricacies of social interactions.

During this whole, protracted period of development, we are entirely reliant on being carried around, fed, kept warm and protected. This places an enormous burden on the mother's time – her ability to gather food and care for the infant, all the while protecting herself and her other offspring – so much so that for our ancestors, raising a baby became hugely challenging for a mother on her own. Thus, as we evolved to be more intelligent, and so increasingly dependent as babies, there was a strong selection pressure for both parents to take an active role in child-rearing.*

---

* This is not to say of course that today, single mothers, or indeed single fathers, aren't able to perform this role admirably. It does require an

PAIR-BONDING

If both parents cooperate and raise their baby together, it has the best possible chance of surviving the vulnerable early years. But each parent needs reassurance that the other is committed to the arrangement. The woman needs to be certain that, in a sexual relationship, the man is likely to stick around through the pregnancy and first years of the baby's life, when she and the child are in most acute need of assistance. In turn, the man has to be confident that he's not being cuckolded – the identity of a child's mother is always clear, but paternity can be much less certain. How can both parties know that their partner is committed to the relationship and any children that may result from it?

The solution that evolution found to this quandary is pair-bonding. If each partner experiences an intense attachment to the other, they will be compelled to cooperate in child-rearing. This human pair-bond is regulated by a hormone called oxytocin.

Oxytocin serves many functions in mammalian reproduction. It stimulates the muscular contractions of the uterus during childbirth and the secretion of milk during lactation but also, crucially, the emotional attachment between a mother and her newborn. All mammalian mothers nurture their young: they suckle them, protect them from predators and teach them vital life skills. Yet experiments have shown that rats that have had the oxytocin messaging blocked after giving birth don't provide any normal care or attention to their pups.[2] On the other hand, virgin female sheep injected with oxytocin will start exhibiting maternal behaviour to a lamb that is not their own.[3]

While this mechanism of mother-offspring bonding is shared by all mammalian species, it has been modified and extended in

---

enormous investment of time and resources, however, and invariably the help of close family or friends, or of wider society and the welfare state, much of which was not available to our ancestors.

humans to create a deep attachment between partners as well. Today, we experience this in the emotion of romantic love. Oxytocin is released in both men and women during sexual intercourse, and especially during orgasm, so that sex helps first to establish and then maintain the pair-bond.[4] The different stages of romantic love – from attraction to attachment – involve the release of another key signalling molecule, called dopamine, in the brain's reward pathway. As we'll explore in Chapter 6, drugs such as caffeine, nicotine and heroin trigger the same pleasure centre in the brain. The neurochemistry of love is very similar to that of addiction – or you could say, love is a drug.[5] Thus evolution has ensured that partners likely to have a baby together are also bound to each other by a mutual hormonal tie.*

If a man sees that the woman loves him, he can be confident that she's not sleeping with anyone else and the children will be his. And in return, if the woman sees that the man loves her, she has a reliable assurance that he will stick around to help raise the baby.[7] In crude evolutionary terms, within a pair-bond, partners are exchanging certainty of paternity for a guarantee of resources.

While long-term pair-bonding is an integral part of human reproduction, this does not mean that faithfulness is always absolute nor that such relationships endure indefinitely. In many pairings, the initial, intense, passionate bond of romantic love mellows before long into a calmer attachment, or eventually falters altogether. The period over which the strength of the attachment begins to fade, until the pair-bond effectively dissolves (at least for one of the two partners), has been found to

---

* While maternal care is the norm among mammals, pair-bonding and male investment in offspring is rare – less than 5 per cent of mammalian species form a stable pair-bond between mates, and less than 10 per cent of mammals show any degree of paternal investment in offspring. In contrast, around 90 per cent of birds form monogamous pairs – mainly just for a single breeding season, but some do form lifelong bonds like the swan or bald eagle. But in all animal species that do form pair-bonds between mating partners to improve reproductive success, it's oxytocin that forges the attachment.[6]

be around four years. Interestingly, this is roughly the time it takes for a child to develop enough that it is no longer critically dependent on the support of both parents.[8] Statistical studies show a peak in the divorce rate between the fourth and sixth year of marriage across many different societies[9] – evidence for the popular conception of the seven-year itch. It seems evolution invented romantic love to ensure the commitment of biparental child-rearing, but only for as long as it's needed for reproductive success.

The emergence of this oxytocin-driven web of ties binding both parents and their offspring together created something very special in our history: the family. Many primate species live in social groups, but we humans are unique among the great apes in sticking together in stable family structures.[10] What's more, as we saw in the previous chapter, humans form strong attachments not only to offspring and sexual partners, but also to wider kin, as well as others to whom we are not related – close friends and the social network. Over our evolutionary history, the range of other individuals humans feel a deep connection to has broadened further and further. And this oxytocin system has even been extended to include other animals, as we domesticated wild wolves and cats to become our pets. We are a species of bonders.

The formation of a pair-bond between sexual partners is essentially a mutual exclusivity contract for reproduction, forged biologically by hormones. Thus the institution of marriage is no more than a social practice built on this evolutionary foundation, formalising the bonding already intrinsic to humans. An analysis of 166 societies around the world concluded that romantic love was a universal feature of the human experience. Formal marriage arrangements between a man and a woman also exist in all known cultures, and 90 per cent of people in the world will marry at least once in their lives.[11] The cultural norms that have built up around marriage, and are codified in religion or law, specify the expectations associated with the union. These include whether the bride or groom lives with the other's family, or whether the newlyweds establish a new household together. They set out rules over inheritance and the division of property following separation or the death of one partner. They may also stipulate a transfer of wealth in the form of a dowry or bride price, so that marriage becomes not only a contract between bride and groom, but a transaction between their families.

But in all of the world's cultures – from the inelaborate betrothal of Hadza hunter-gatherers in north Tanzania to the chant- and ritual-filled Greek Orthodox ceremony to the three-day celebration of a Hindu wedding – at its core, marriage is a public declaration of a couple's mutually confirmed commitment (or, in the case of a polygamous marriage, commitment between multiple spouses). Wedding customs are culturally specific and have changed over time, but the practice of partners publicly committing to each other in order to regulate reproductive behaviour must be as old as human language, if not older.

Living within family groups, and supporting our close kin, has been a central part of human existence from the beginning of our species. Family has taken a variety of forms, from an extended family of relatives, spanning several generations, living together in the same household, to the nuclear family of a single couple with their children that became common in the post-industrial west (and of which the single-parent family is a variation).[12] For much of history, in the absence of state institutions, one's family was the only resource available to support a person during sickness or old age.

The development of agriculture and the emergence of civilisation introduced another key aspect of family life. When our hunter-gatherer ancestors took to farming and abandoned their mobile lifestyle, the ability to accumulate possessions increased markedly – whether that was pottery, metal tools, a flock of goats or a hoard of precious metal currency. Agriculture also created the concept of land ownership – proprietary rights over a particular tract that is tended and tilled by the same family (or its serfs) to grow crops or pasture livestock.

These possessions could be passed from parents to children, keeping their benefits within the same family (in an extension of the kin selection we discussed in the previous chapter).* We had always inherited physical traits from our parents, but now material wealth could also be transmitted from one generation to the next. And it wasn't just the assets or land themselves, but also the influence and status that they afforded. For those at the top of society, control over a whole territory and the people and resources within it was also inherited. Steep social hierarchies and levels of inequality developed and perpetuated what would have been unknown to our hunter-gatherer ancestors – rich vs. poor, rulers vs. ruled.

The most prominent position was occupied by the king, the sovereign ruler over an entire state. Those kings able to muster the larger and more formidable armies conquered other chiefdoms which they incorporated into their realm. Over time, patchworks of independent territories fell under the hegemony of a single supreme ruler – a king-of-kings, an emperor.†

While children within the elite social strata inherited the wealth and status of their parents, family professions were also passed down within other sectors of society. By being exposed

---

* Particularly treasured items passed intact for many generations – the original meaning of an 'heirloom' was a valuable tool or utensil passed to an heir.[13]

† Several modern names for such supreme rulers are directly derived from the title Caesar and were employed to invoke the imperial grandeur and conquest of the ancient Roman Empire. The medieval Holy Roman Emperors called themselves Kaiser from the tenth century, and Ivan the Terrible adopted the title of Tsar in 1547;[14] they all considered themselves successors of the ancient Roman Emperors. Mehmed II, and then subsequent Ottoman Sultans, took the title Kayser-i Rum meaning 'Caesar of Rome', in reference to Mehmed's conquest of the Byzantine Empire in 1453. The use of Caesar as a title for Roman Emperors was itself derived from the nickname for Gaius Julius, the general and statesman who played a key role in the end of the Roman Republic in the first century BC. Referring to his baldness, contemporaries jokingly called Julius 'Caesar', meaning hairy. This nickname has persisted through history to be adopted proudly by emperors across Eurasia.[15]

to, and learning, the skills from a young age, sons often stepped into their father's trade and role within society, inheriting the necessary tools: baker, butcher, miller, mason, sawyer, wright, smith. Indeed, in medieval England, for example, many of these common family trades became adopted as surnames.[16]

Inheritance customs and laws have varied between societies but are often distinguished between realty (land and buildings) and personalty (such as household goods, personal effects, live-stock and cash).[17] The central quandary facing any testator was: what's the best strategy of inheritance to ensure future success? The fairest way for wealth and land to be passed down within a family might be partible inheritance – the sharing of posses-sions equally among all children, or at least all sons.* But for land inheritance the problem is that with each generation the original tract becomes parcelled down into ever-smaller subdi-visions, until eventually each is no longer sufficiently productive. For the aristocracy, this dispersal of territory also represents a disintegration of wealth and influence.

An alternative system is to pass the majority of the family realty to a single heir. Primogeniture – inheritance by the eldest son – began to be adopted in medieval Europe among the feudal nobility (and later among landholding peasants too) and else-where around the world. Primogeniture prevents the splitting of estates and the titles and privileges that went with them.

The favouring of the firstborn son with the family's estate meant that younger sons were forced to seek careers within military service or the Church. Primogeniture is also thought to have been a major factor behind the Viking Age. From the end of the eighth to the mid-eleventh century, fierce sailors erupted

---

* Patriarchy, and the concentration of power and privilege among men, is seen in most societies.[18] While there are biological differences between the sexes, there is no clear anthropological reason why sex inequality and patri-archy became common in societies. It's also worth stressing that patriarchal societies can still observe matrilineal customs, with inheritance of property and title down the female line. Matrilineal inheritance is found, for example, within some parts of Africa, southeast Asia, and pre-Columbian America.[19]

out of Denmark, Norway and Sweden and across Northern Europe aboard their longships. After an initial fifty years of smash-and-grab raids on vulnerable coastal monasteries in the British Isles, the Scandinavians increasingly took to settling in the regions they invaded – a shift believed to have been driven by an increase in the number of younger sons in Scandinavia with no chance of obtaining the family's holdings and so being forced to venture abroad to secure farmland of their own. This expansion eventually produced Viking settlements in England, north Scotland and southern Ireland, as well as in the Baltic, Russia and Normandy (William the Conqueror was himself a descendant of the Viking leader Rollo).[20]

Many of the conquistadors leading expeditions into Central and South America in the sixteenth century were also younger sons of the nobility who stood to inherit little from their wealthy families.[21] Similarly, many of the settlers sailing to North America – and particularly to the plantations of the southern colonies – in the eighteenth century were the younger sons of British gentry, who stood to inherit no land of their own at home but did receive some money to establish themselves elsewhere.

Primogeniture was first rejected in favour of partible inheritance by the New England colonies in the mid-seventeenth century and then abolished by law across the United States shortly after the declaration of independence.[22] Within twenty years, the French Revolution had also abolished primogeniture with its sweeping away of the Ancien Régime. Elsewhere, primogeniture declined as societies progressed through the demographic transition (which we'll discuss in Chapter 5); families began to have fewer children and ownership of land became less central to economic success.[23]

Today, most nation states are run by representational democratic governments – although with varying degrees of corruption and political repression[24] – which accede to power through election. Our leaders are expected to display capability and merit, while nepotism in public life is frowned upon. But this is a relatively recent state of affairs: for millennia, since the earliest emergence of civilisation, absolute rulers have reigned supreme. And while power lay in the hands of a single individual, it was usually inherited within the same family: kingship as intimately tied to kinship.[25] The result of this fusion of family lineage and inheritable status is dynasty – an extended family that passes wealth, territory and power from one generation to the next. A dynasty behaves like a superorganism, preoccupied with its own survival and advancement, striving to cling on to and expand its territory, prestige and influence while competing with other families or intermarrying to further their own interests.

Dynasties became such a prevalent feature of human civilisation that their names often serve as a convenient means for delineating historical periods of a particular state, empire or region: we talk about the Tudor age, the Ming dynasty, Tokugawa-period Japan and so on. These names are shorthand and encompass not just the ruling family but the major cultural, socioeconomic, military or technological trends or events dominant at that time.

So although marriage is a universal social construct built upon the human predisposition to form pair-bonds, within dynasties marriage took on a whole new importance. Far more than just the union of two individuals, it represented the tying together of two powerful families; and strategic marriages were meant to cement political alliances. The children born into these marriages intertwined the bloodlines of both dynasties, literally embodying the accord between the two powerful families. The human imperatives of pair-bonding and reproduction became tools of statecraft.

For a royal family, births, deaths and marriages are all political events, and they have profound repercussions for all those living within the kingdom or empire. At the same time, such family dynamics dominate international relations as well. In European history, the kings of one particular dynasty were masters of this grand design.

THE HABSBURGS

Asked to name a great royal family in history, we might think of Charlemagne's medieval Carolingian dynasty, the House of Bourbon in France or England's Tudors. But none have had as great an impact across the breadth of Europe, and globally, as the Habsburgs. As the predominant royal family for around half a millennium, they acquired a vast empire across the continent and around the world. The gradual but persistent growth of Habsburg territories happened, on the whole, not through sweeping military conquest but by accumulating crowns in a carefully orchestrated programme of strategic royal marriages.[26]

Rising from humble origins in Swabia, in today's northern Switzerland, the Habsburg family's fortunes experienced a significant boost in the mid-fifteenth century when they were able to manoeuvre themselves into a favourable position within the college of princes who elect the Holy Roman Emperor. The Holy Roman Empire had been founded in 800 when Charlemagne, King of the Franks, was crowned by the Pope as emperor of the Romans, supposedly in continuation of the ancient Roman Empire. From the tenth century it became a largely Germanic empire but included many other kingdoms through Central Europe and down to the Mediterranean and Baltic coastlines.[27] Although emperor was formally an elected post, the incumbent was often able to wield sufficient influence to ensure his own son was chosen as successor, in what became a de facto hereditary imperial crown. For three centuries between 1438 and 1740, all the Holy Roman Emperors were Habsburgs,[28] and the dynasty was able to merge this conglomerate of Central European kingdoms with its own territories.

The architect of the Habsburg's phenomenal ascendancy in the fifteenth century was Maximilian I, who actively pursued a policy of dynastic marriages between his family and other prominent royal houses in Europe.[29] In 1477, he married the heiress of the Duchy of Burgundy, acquiring not only this territory on the eastern boundary of France, but also the Low Countries (Luxembourg, Belgium and the Netherlands), enabling him to tap the wealth moving through their ports. Maximilian then arranged for his son Philip to marry Juana, the daughter of Isabella of Castile and Ferdinand of Aragon, in 1496. Following the deaths of her elder siblings and nephew, Juana inherited both crowns, and her and Philip's son, Charles, became the king of a unified Spain.* With this Spanish inheritance the Habsburgs also came into possession of southern Italy, Sardinia and Sicily, as well as settlements along the north African coastline.[31] It turned out to be fortuitous timing. Just four years before Philip and Juana's wedding, Columbus had crossed the Atlantic and discovered the 'New World'. The Spanish branch of the Habsburg dynasty thus came to rule over a dominion extending far beyond the European peninsula, as conquistadors and colonisers laid claim to vast tracts of territory across the Americas. Then, in 1521, the naval explorer

---

* This fusion of the kingdoms of Castile and Aragon created Spain as we know it today. Other prominent features of the modern political map of Europe were created when the opposite happened and an empire was divided among heirs. At its greatest extent in the early ninth century under Charlemagne, the Carolingian empire stretched across what is today France, the Low Countries, northern Italy, Austria and Germany. After his death and that of his son Louis I, Charlemagne's three grandsons fought a vicious civil war over their inheritance. After three years of bloodshed, the brothers agreed at the Treaty of Verdun in 843 to partition the empire among themselves: the Kingdom of the West Franks came to form the basis of modern-day France; the Kingdom of the East Franks became the Holy Roman Empire and later Germany; and an elongated middle kingdom reached from northern Italy to the Low Countries on the North Sea. By the early tenth century, the northern territories of this middle kingdom had been mostly absorbed into East Francia, and so the division of Charlemagne's empire decided by three squabbling brothers became permanently engraved in the map as the border between what would become the two most powerful nations of continental Europe, France and Germany. In the great wars of the twentieth century, millions of young men died fighting over a line drawn as a family compromise over a millennium previously.[30]

Magellan claimed the Philippine islands for Spain (named after Charles' son, Philip II).[32]

Yet Maximilian's ambitions did not stop there. He arranged for his grandson Charles to marry Isabella of Portugal in 1526. This completed the absorption of the whole Iberian Peninsula into the Habsburg realm, as well as the Portuguese conquests in Brazil, India and the Spice Islands. Maximilian also arranged for his other grandson, Charles' younger brother Ferdinand, to marry into the Hungarian royalty. When the Hungarian king was killed in battle against the Ottomans in 1526, leaving no heirs of his own, the crowns of Hungary, Bohemia and Croatia also passed into the Habsburg family, delivering the territory that would form the core part of their Central European empire for the next four centuries.[33]*

The Habsburgs had transformed themselves from middle-ranking counts in Swabia to the principal dynasty in Europe. And within just 50 years of prudently plotted marriages, they had acquired over half of the continent;[35] what's more, they had achieved this more or less bloodlessly. As the seventeenth-century saying went: 'While others wage war, you happy Austria marry!'[36] While they had had to defend some of their claims through force of arms (and although Spanish and Portuguese invasions in the New World and South East Asia were brutal), the majority of this astonishing growth in influence had come about through strategic royal unions and the gradual accumulation of crowns and territories through inheritance.[37] The Habsburgs were true grandmasters of the game of thrones.†

---

* For a brief period, there was even a Habsburg emplaced on the throne of England. Holy Roman Emperor Charles V organised the second marriage of his son, Philip, to Queen Mary I in 1554, making him *jure uxoris* ('by right of his wife') King of England and Ireland.[34] But when Bloody Mary died four years later, the crown passed to her half-sister, Elizabeth I, returning the throne to Protestantism.

† In terms of the growth of its European territory, the Habsburg family has only been challenged in recent history by Napoleon in the nineteenth century and Hitler in the twentieth,[38] but these empires seized by blitzkrieg war were

They also benefitted from a good degree of biological luck. In the system of inheritance and succession, royal families that fail to produce any surviving children, or at least any male heirs, risk having their titles and territories claimed by more distant relatives or in-laws. For centuries, the Habsburgs reliably produced male heirs, or at least surviving nephews and male cousins, through whom they could lay a claim to the kingdoms they had married into which had faltered in the male line. By way of this genealogical endurance, by simply out-surviving rival families, the Habsburgs were able to absorb their territories and wealth – we might call it the 'last man standing' approach to territorial expansion.[39] The historian Martyn Rady calls it 'the Fortinbras effect', after the Prince of Norway in Shakespeare's *Hamlet*, who arrives at the end of the play to find all his rivals dead and so claims the vacant throne.[40]

By the mid-sixteenth century, the Habsburgs had manoeuvred themselves to become a dominant power not just within Europe but across the Atlantic and Pacific Oceans – the arms of this single family encircling the world.[41] And this vast dominion was all ruled by one man, Charles V, Holy Roman Emperor, the first ruler in history to reign over an empire on which the sun never set. But he knew that he couldn't pass this all on to either his brother or his son, as neither would quietly concede their claim to the other, and so at the point of its greatest extent, the Habsburg empire was cleaved in two. Maximilian's grandsons initiated two branches of the Habsburg dynasty: Charles passed the lands of the Spanish Habsburgs (including the Low Countries and the worldwide Spanish possessions) to his son, Philip II; while Charles' brother Ferdinand I received the ancestral lands in Austria, and his Central European descendants continued as rulers of the Holy Roman Empire.[42]

At the beginning of the eighteenth century, Habsburg global power suffered a sharp knock with the loss of the Spanish

---

held for just a few years – vanishingly ephemeral when compared to the territories the Habsburg family consolidated over generations.

branch, but the Central European dynasty remained a considerable European power, morphing into the Austro-Hungarian Empire in 1867. The family was still an instrumental player in the affairs of the twentieth century when, on 28 June 1914, the heir presumptive, Archduke Franz Ferdinand, was assassinated in Sarejevo. Within little more than a month, the world was being consumed by the most devastating war it had yet seen. Defeat in the First World War delivered the final blow to Habsburg power with the collapse of the Austro-Hungarian Empire and the dynasty losing all its remaining territories. (The family itself still survives: Karl von Habsburg, grandson of the last Austro-Hungarian emperor, Charles I, serves today as an Austrian politician.) But for over four hundred years, this single extended family had been a principal player in European and global affairs.

MONOGAMY VS POLYGAMY

The Habsburg dynasty, and indeed all of Europe and its colonies around the world, lived with a cultural norm of monogamy. But polygamy has been extremely prevalent through world history. An ethnographic survey of 849 human cultures – including both hunter-gatherer groups and agricultural societies – found that 83 per cent of them are polygamous, with virtually all of these containing polygyny rather than polyandry.[43] These similar terms can get a little confusing. Polygamy is the general term for having multiple spouses. Polygyny is the form of polygamy where a man has multiple wives, and polyandry is a woman having more than one husband. Polygyny is far more common than polyandry, but while it is permitted in many cultures, it is important to note that even in these societies it is usually only the highest-status men who are able to support multiple wives, and so the majority of men and women in those societies still live in monogamous pairings.[44]

Polyandry is found in less than 1 per cent of societies in

anthropological records.[45] While there have been examples of powerful queens taking many husbands – such as Nzinga, a mid-seventeenth-century warrior-queen of Ndongo and Matamba (in today's Angola)[46] – polyandry typically involves brothers marrying the same woman, often in response to environmental circumstances.[47] For example, in the Tibetan highlands and foot-hills of northern India, the harsh landscape makes it difficult to cultivate enough food to support a family, so when a plot of land is partitioned too often between family members it becomes insufficient. With fraternal polyandry, a group of brothers mar-ries the same woman, and they work together to farm their land so it doesn't become subdivided in each generation among sepa-rate families.[48] It is therefore a different solution to the same problem that primogeniture addresses. Several men married to one woman also produces slow, sustainable population growth, so polyandry offers a demographic solution to ecological constraints.

Monogamy is believed to have been the predominant condi-tion for our hunter-gatherer ancestors. Today's Hazda people, living in the savannah woodlands of northern Tanzania, repre-sent a fairly typical example of the hunter-gatherer lifestyle and so offer insights into how humanity probably lived tens of thousands of years ago. The Hazda form small foraging bands of around 30 individuals, who move camp about every two months as they exhaust the available food in one area. These groups are very fluid, with individuals drifting between nearby camps or whole groups splitting or merging. All food brought back is shared among everyone in the camp. With no means of preserving food, the Hazda aren't able to save reserves for later or build up a surplus; their mobile lifestyle also means they keep few material possessions and carry only what they need to survive. Theirs is a remarkably egalitarian society, without stark resource disparity or significant hierarchy between adults, and with equality between men and women. Monogamy is the social norm among the Hazda, with very few men having two wives at the same time; usually, one of the women becomes unhappy in a polygynous relationship and leaves. As the women are generally able to gather enough food for themselves, or share with everyone else what is brought back to the camp, their self-sufficiency frees them to divorce their husband.[49] Anthropologists assume that our hunter-gatherer ancestors would have led similar lives.

As we have seen, with the emergence of agriculture, individuals started to accumulate wealth and status, and highly hierarchical societies developed. Men at the top of the social pyramid were able to support several wives, and polygyny became the widespread norm, rather than the exception.[50] The Kaguru, for example, agrarian people that live south-east of the Hazda in today's Tanzania, practise polygyny.[51] Polygyny was also the norm for the ruling elites in Asia and the pre-Columbian Americas.[52] Poor men, on the other hand, have been monogamists all round the world.[53]

While polygyny is obviously advantageous for the reproductive potential of powerful men, wives can also benefit from the arrangement. In an egalitarian society with roughly uniform distribution of resources among individuals, such as a hunter-gatherer community, females benefit most from the undivided attention of one man for co-parenting their investment-intensive offspring, rather than sharing that man with other females. But if there is greater disparity in males in terms of their status, wealth or other ability to provide, it may be in the woman's interest to have a smaller share of the ample resources of a wealthy, high-status male, rather than be the sole mate of a resource-poor male.[54]

The anomalous pattern of monogamy in the West today originated in the ancient civilisations of the Mediterranean. In efforts to foster a more egalitarian and democratic society – by offering all male citizens the chance to find a wife – from around 1000 to 600 BC, the Greek city states enacted laws regarding monogamy.[55] These cultural norms were later adopted by the Romans and expanded with the introduction of new laws to restrict polygamy and strengthen the institution of monogamous marriage.

Between 18 BC and AD 9, for example, Emperor Augustus, concerned with moral decay and political decline, restricted the inheritance unmarried men would be eligible to receive and legally formalised the process of divorce so as to discourage serial monogamy. Furthermore, married men were prohibited from taking concubines – although, they were permitted to have extramarital sex with prostitutes and to exploit slaves, who often bore illegitimate children.[56] While polygyny was officially viewed by the ancient Greeks and then Romans as a degenerate custom of uncivilised barbarians and monogamy as the social and legal norm, in reality many men practised a form of de facto sexual polygyny.[57]

The militaristic expansion of the Roman Empire forced monogamy upon much of Europe, and after the collapse of the Western Roman Empire, the Christian Church continued to promote this cultural norm.[58] There is nothing inherently monogamous in the Judaeo-Christian tradition. The patriarchs and kings of the Old Testament had several wives – most notably King Solomon who is said to have had 700 wives and 300 concubines.[59] Nor did the rich and powerful men of Christian Europe remain strictly monogamous; they often had one wife, who bore them legitimate heirs, but also a number of mistresses or concubines. With the colonial expansion of Christian European states from the early sixteenth century, their cultural norms and legal systems around monogamy were exported around the world and imposed upon indigenous societies. Today, monogamous marriage prevails around the world – but polygamy is legal in 28 per cent of sovereign states, mostly Muslim-majority countries across North Africa, the Arabian Peninsula and Southern Asia.[60]

It is clear, therefore, that humanity has a predilection for polygamy but in our ancestral condition, this was kept in check by the resources available in a hunter-gatherer community, and monogamy prevailed. The social inequality brought about by agriculture, however, allowed polygynous impulses to be manifested by powerful men. In Europe, legal systems developed to reinstate monogamy as the cultural norm, which was then also enforced across the world by colonialism. Indeed, if people didn't already have a proclivity towards entering into polygamous marriages, there would have been no necessity to outlaw it. Kings and emperors through history permitted themselves to support many wives or maintain large harems – as we have seen, polygamy is a consequence of social hierarchy. Wealth disparity is as large today as it has ever been, with a vast gulf between the ultra-rich industrialists and internet entrepreneurs and those living in destitution. Were it not for the prevailing cultural norms, reinforced by top-down laws, and the US had followed the pattern seen in many polygynous societies, it could be the most extreme polygynous state in history. At the time of writing, Elon Musk has a personal net worth of nearly a quarter of a trillion dollars. With this fortune he could materially support hundreds of thousands of wives, utterly overshadowing the harems of the greatest despots of history.

DYNASTIC REPRODUCTION

Human reproductive behaviour is a rich, mixed repertoire, and in our history, we have combined the promiscuity of chimpanzees, the monogamy of gibbons and the polygyny of gorillas. We undeniably display a predilection for polygamy, but monogamy became our predominant ancestral condition – and has been culturally and legally enforced in Europe for centuries – even if wealthy, high-status individuals in many societies and civilisations have practised polygyny. The consequences of these two reproductive systems have had a profound influence on world history, particularly when it comes to how political power was passed on from one generation to the next.

In late-medieval Europe, most royal families (as well as those of the aristocracy) adopted primogeniture – sole inheritance of the kingdom by the eldest son. The successor states of the Carolingian Empire, which extended across France, the Low

Countries, northern Italy, Austria and Germany, went one step further and expressly forbade not only a woman to ascend to the throne but even inheritance to be transferred through the female line of descent – known as the Salic law.[61] What's more, European monarchies often excluded illegitimate sons from the succession – those fathered by the king but not born of his queen. In this way, the right to rule was defined not by the royal bloodline transferred from the king, but by the marital status of the king and mother of the child.[62]

The advantage of such clearly established rules of succession was that they minimised uncertainty over the rightful heir and the number of rival claimants to the throne on the king's death, and so offered a dynasty a smooth transfer of power and a stable state.[63] As the French political philosopher Montesquieu wrote in 'The Spirit of the Laws' in 1748, 'The order of succession is not fixed for the sake of the reigning family; but because it is the interest of the state that it should have a reigning family.'[64] The heir apparent could also be prepared for his future role as absolute ruler. But while a strict order of succession offers clarity, a system of pedigree over merit risks producing weak or incompetent kings: the person most suitable for leadership is unlikely to just happen to be the eldest son of the previous ruler (especially if the heir apparent is still a young child, or infirm).[65]

Another problem of primogeniture within the monogamous tradition is that it puts pressure on the king to ensure he has at least one surviving legitimate male heir before he dies. The urge to reproduce is an instinctual drive within humanity, as it is in any species, but in a family transferring sovereign power, continuation of the bloodline takes on an even greater significance. It is the king's duty to produce 'an heir and a spare' (in case of the premature death of the eldest) to continue the legacy and maintain the prestige of his dynasty. Failure in this act has major repercussions – a succession crisis can lead to civil war affecting the entire population, and power might pass to another family, or even a rival kingdom.

Yet within a monogamous system in particular, the creation of heirs is limited by human reproductive biology. A queen can carry only one child at a time – discounting twins, which account for only a few per cent of pregnancies – and with high child mortality before the modern era, there was a significant chance that a king may die leaving only daughters or no surviving children at all. Then the future of the entire royal dynasty is at stake. Desperation over producing an heir was a main reason Henry VIII felt compelled to divorce or behead one wife after another. You could say, therefore, that his separation of the Church of England from the authority of the Pope was an effort to permit serial monogamy – functionally equivalent to polygyny but spread over time.[66]*

Within societies that permit polygyny, however, there is no such anxiety over a single queen bearing an heir. Henry VIII's contemporary, Suleiman the Magnificent, sultan of the Ottoman Empire, had more than a dozen 'legitimate' children by his harem.[67] And imperial harems often produced many more children than this. While a woman cannot typically produce more than about a dozen surviving offspring in her reproductive lifetime,†[68] a king with access to a harem of women can achieve an extraordinary fecundity.‡ The early-twelfth-century emperor Huizong holds the record for China: 65 children (including 31

---

* Before Henry's father ascended the throne and established the House of Tudor, the kingdom had been wracked by thirty years of civil war – the Wars of the Roses – over rival claims to the crown, and Henry was keen to avoid a similar succession crisis.

† The greatest number of offspring reliably recorded for a single woman is 69, born to a Russian woman living in the eighteenth century, but this is very much an outlier.

‡ The Arabic word *harim* refers to the part of a house where access is forbidden. It is an inner sanctum and the private chambers of the emperor, out of bounds to the palace's administrators, magistrates and visiting dignitaries. Its purpose was not solely to house a large group of women[69] available for the emperor's sexual gratification. It also housed wet nurses for the numerous royal children, as well as the ruler's female relatives and their maids.[70]

sons who were mothered by twelve consorts) are named in the
Song Shi, the official history of the Song Dynasty, although
there were probably more – he was brought a new virgin at
least once a week.[71] The Ottoman sultan Murad III was survived by 49 children, and another seven of his concubines were
pregnant when he died at the end of the sixteenth century.[72] The
late-eighteenth-century shogun Tokugawa Ienari holds the Japanese record with 52 children from his 41 concubines and
sixteen consorts.[73]*

Royal families that reproduced polygynously rarely feared
for their dynastic continuity, but without clear succession rules
(such as primogeniture) in place, competition between potential
heirs could quickly turn murderous.[76] Within the early Ottoman Empire, for example, the death of the incumbent ruler
served as the starting gun for a violent contest between the
princes – a true Battle Royale – with the victor either killing all
his rivals in battle or committing fratricide after securing the
throne.[77] When Mehmed III claimed the crown in 1595, for
instance, he had all nineteen of his brothers murdered, along
with every pregnant woman in his father's harem. Pruning the
royal family tree of contenders removed the risk of challenges
to the throne. Such unsettled patterns of succession and the
ensuing contests between potential heirs often created unstable
interregnum periods after the death of each ruler; but a competitive scramble for the throne does at least constitute an
exacting selective process. The prince who is able to rally the
greatest support behind his own claim, or who demonstrates

* These huge numbers of children sired by emperors have left long-lasting
genetic footprints in the population. One particular association of genes on
the Y chromosome (and so only present within males) that is found in around
3 per cent of all East Asian men living today is believed to have descended
from the dynasty of Qing emperors who ruled China from the mid-seventeenth
century.[74] Another Y chromosome linkage found in about 8 per cent of men
across a huge tract of Asia – and up to 1 in 200 men worldwide – is believed
to have originated with Genghis Khan and his brothers, who surged across
the continent in the thirteenth century to forge the Mongol Empire.[75]

the most cunning and courage on the battlefield, has demonstrated the sort of qualities also needed in a supreme ruler.[78]

By the early seventeenth century, the Ottomans settled on a much less bloody solution to the biological problem of reproduction within the dynasty. The reigning sultan confined all his male relatives within the walls of the palace (a practice also adopted by the Mughal and Safavid dynasties in India and Iran respectively),[79] where they lived comfortably but were forbidden from having children of their own[80] – a true gilded cage. Agnatic seniority was also adopted, whereby the sultanship passed to the sultan's eldest younger brother, and upon his death the next eldest, and only after all the brothers have died, to the eldest son of the first. Only the sultan was able to produce heirs – exerting reproductive control to preserve the stability of succession.[81] As a result of the harem system, the Ottomans achieved a remarkable dynastic longevity – an unbroken chain of each sultan being succeeded by either his own son or the son of one of his predecessor-brothers, stretching over 600 years.[82]

Harems enabled rulers to maximise their own reproductive potential – and indeed, possession of a large harem was itself a status symbol – and so the Ottoman sultan in Topkapı Palace and the Chinese emperor in the Forbidden Palace, amid all their opulence and wealth, were no different in their fundamental biological motivation from a silverback gorilla sitting among his female mates and beating his chest to ward off rivals.

The operation of a harem for royal reproduction posed a potential problem, however. In polygynous animal species such as gorillas or lions, the dominant male keeps a careful eye over his harem of mates, fending off any potential rivals. In an imperial court, an emperor could not personally watch his harem, so he posted guards in the inner court. But how could he be sure that none of the guards would be tempted to have sex with the women himself? How could he be absolutely certain of the paternity of his children, and crucially his potential heirs, born of the harem? He could post several guards to watch each other, but that doesn't stop them potentially colluding. The solution was to select only infertile men, or indeed men that had been made sterile: eunuchs. The creation of a sterile caste by castration – the removal of both testicles, and sometimes the penis as well – extends back to the first civilisations,[83] but the word we use in English today, 'eunuch', derives from the Greek εὐνοῦχος, meaning 'bedkeeper', relating to their role as personal servants. The practice probably began with slaves who had little choice over the matter (as little as many of the women within the harem), but in China, eunuchs came to hold a revered position, and men volunteered for the honour.[84]

Such men-without-manhood came to serve many roles in imperial courts, not just as wardens of the harem but also as personal assistants, royal bodyguards and palace administrators, and beyond the palace as provincial governors, soldiers and military commanders.[85] The fact that they were sterile, and often also forbidden from marrying, meant they had no investment in their own legacy or family loyalty.[86] Eunuchs were therefore less likely to have ulterior motives, and so were believed to be more devoted and trustworthy servants of the royal court. Those in the inner sanctum of the harem occupied positions with unique access to the emperor, serving as confidant and counsel. And in palaces with a strict divide between inner and outer courts, the emperor didn't communicate directly with his chief advisors; eunuchs would also serve as intermediaries.

By the tenth century in the Byzantine Empire, half of the administrative ranks in Constantinople were reserved solely for eunuchs, who often outranked the 'bearded' civil servants.[87] In China, the sterile caste numbered around 10,000 in and around the Forbidden City in the 1520s, and this expanded over time so that by the end of the Ming dynasty in the early seventeenth century, the administration in Beijing contained a staggering 70,000 eunuchs, with another 30,000 distributed across the empire as government administrators and provincial governors.[88]

These imperial courts of history can be seen to resemble the hives of eusocial insects such as bees or ants. The emperor in the inner sanctum of the palace predominated over reproduction with a harem of mates and was surrounded by members of a sterile caste performing key roles both within the household itself and across the imperial dominion, as attendants, guards, administrators and military commanders. The gender roles are reversed, however: eusocial insects have a single queen at the centre of the colony, reproducing with a group of male drones and attended to by sterile female workers.

Let's return now to the European monarchies operating within the cultural norm of monogamy. For those dynasties that tried to keep power within a close family, there was a biological sting in the tail.

CURSE OF THE SPANISH HABSBURGS

We saw earlier how the Habsburg dynasty had once been the master of strategic marriage, deftly constructing a web of family connections with the other great houses of Europe. The Habsburgs had also been biologically fortuitous, not only consistently producing male heirs (or at least numerous nephews and male cousins with strong succession claims) to secure the continuation of their own line but also often outliving rival families they'd married and inheriting their territories. But then the Habsburgs began to falter. You would expect a dynastic hiccup sooner or later simply from the balance of probabilities – a king dying young before he had been able to father an heir, a king or his wife being infertile – but the Habsburgs were unwittingly stacking the genetic deck against themselves.

While marriages into other ruling dynasties initially spread their influence, in order to prevent their prodigious political authority diffusing away and keep their empire intact they repeatedly married their own close relatives – cousin and cousin, or uncle and niece – especially for the line of kings (and far more so than did other royal houses). Yet while these consanguineous marriages reinforced their political power, such inbreeding consolidated defective genes within the family. Over the generations, the Habsburgs inflicted upon themselves a greater and greater hereditary burden. The means for their ascendency, therefore, also held the seeds of the catastrophic downfall of the Spanish Habsburgs, the branch of the family descending though Philip II.

The problem is genetic variation. When a child is conceived, it receives two copies of each gene, one from its mother's egg and one from its father's sperm. Sometimes, these gene copies, known as alleles, are defective – they have mutated and now produce a protein in the body that doesn't function properly. But since mutations are rare, when a child does inherit a defective allele, the second allele is usually normal and thus able to compensate for the other. These hidden genetic abnormalities are called recessive deleterious mutations. Parents that are closely related to each other, however, will likely already share many of the same gene variants, and so the chances of the child being dealt two defective copies of the same gene are much higher. The effects of the mutation are no longer masked and become manifest as genetic disorders or congenital defects.

Consanguineous pairings create overlaps in the family tree, with certain individuals playing what would be two roles in a typical family structure. For example, in a marriage of first cousins, the children have only three different sets of great-grandparents rather than four. This common ancestry means fewer alleles being contributed to the child's genetic mix, increasing the chances of two defective alleles of the same gene being paired up and causing problems. The probability that the child inherits two of the same alleles of any gene because of this common ancestry is one-sixteenth (0.0625), and so the inbreeding coefficient is 0.0625.* This decrease in genetic variation is

---

* The closest two parents can be related is by sharing half of their genes with each other, as siblings do with each other, or children with their parents. So the greatest degree of inbreeding that is possible within a single generation – resulting from a brother-sister or parent-child pairing – is 0.25. Such close

less significant with pairings of less closely related individuals, such as between second cousins, but if such consanguineous unions are repeated generation after generation, the inbreeding coefficient still mounts greatly.*

An ideal family tree for a child, with good outbreeding (the opposite of inbreeding), looks like a neat branching diagram with eight great-grandparents at the top. But the Habsburg genealogy came to resemble a tangled bush with branches crossing over and even fusing together (in cases where an uncle married his own niece). Out of the 73 marriages entered into by both the Spanish and the Central European branches of the Habsburg dynasty until 1750, there were four uncle-niece pairings, eleven between first cousins, four between first cousins once removed, eight between second cousins and many others among family members more distantly related. Marriages between close kin were especially common among the Spanish Habsburgs: out of a total of eleven marriages in the line of kings, nine were consanguineous unions (third cousins or closer), including two uncle-niece and one first cousin marriage.[91] This increased the degree of inbreeding tenfold over the two centuries separating the founding of this branch of the family, with Philip II, to Charles II, the last Habsburg king of Spain, who had an inbreeding coefficient of 0.254 – even greater than

---

pairings are rare within history, but incest did occur repeatedly within the Incan royal family, as well as in the Egyptian pharaoh's line between the third and second millennia BC, and again within the Ptolemaic dynasty after 210 BC. The mummy of Ahmose Nefertari, the daughter of full siblings in around 1550 BC, shows a conspicuously protruding jaw, reminiscent of the Habsburg jaw which we'll come back to.[89]

* The optimum degree of genetic similarity for maximising the number of surviving offspring appears to be between partners genetically equivalent to third or fourth cousins. For couples more closely related than this, the disadvantages of inbreeding begin to come into play. And having children with a partner who is too genetically dissimilar may break up clusters of genes that have co-adapted to work well together.[90]

the progeny of a directly incestuous parent-child or brother-sister pairing.*

The most conspicuous trait of the dynasty was evident in the distinctive Habsburg countenance. Already prominent with Holy Roman Emperor Charles V, in the early 1500s, but becoming more and more extreme over the following generations, was a long, humped nose with an overhanging tip, and a drooping, bulbous lower lip.[92] Leopold I, reigning as the Holy Roman Emperor in the second half of the seventeenth century, was referred to by the Viennese as 'Fotzenpoidl' on account of his grotesquely swollen lips.[93] (*Fotze* is a particularly crude German word for vagina, and so 'Fotzenpoidl' could be translated as 'twat-face'.) But, in particular, members of the dynasty became characterised by their sharply jutting lower jaw, which was so pronounced that the upper and lower rows of teeth didn't meet; it became known as the Habsburg jaw.[94]

Facial surgeons have analysed portraits of the Habsburg dynasty, focussing on 66 paintings where it can be ascertained that the artist had personally seen the subject, meaning the depiction can be considered reliable, and rated the degree of deformity of the mandible bone of the lower jaw. Comparing these ratings against the calculated degrees of inbreeding for the members of the dynasty confirmed that the protruding Habsburg jaw is indeed correlated with increasing inbreeding, and that it is caused by the effects of recessive genes.[95]

But a deformed jaw was not the only affliction of the Habsburgs. The dynasty came increasingly to suffer epilepsy and other mental issues, generally sickly children, and strings of miscarriages and still-births.[96] Of the 34 children born into the Spanish branch of the Habsburgs between Charles V and Charles II, ten of them died within their first year and a further

---

* The naming of monarchs can get very confusing. The Habsburg king Charles II lived 150 years after Charles V (his great-great-grandfather), and the numbering relates to the realm over which they ruled: Charles II of Spain and Charles V, Holy Roman Emperor (who was also Charles I of Spain).

The architect of the Habsburg's network of strategic royal marriages from
the late fifteenth century, Maximilian I, Holy Roman Emperor (painted
1508); and his great-great-great-great-grandson Charles II of Spain (painted
1685) showing the pronounced 'Habsburg jaw'.

seventeen before their tenth birthday – an overall infant mortality of 80 per cent.[97] Here was one of the most privileged and
pampered families on earth, with access to the best nutrition,
lifestyle and medical attention available at the time, yet they
were suffering child death rates four times higher than Spanish
peasant families living in rural villages.[98] Of those who survived, many were inflicted with not just the infamous drooping
lip and jutting Habsburg jaw, but a range of other physical
deformities.

Matters came to a head with the ascent of Charles II in 1665.
So severe were his many afflictions that he came to be known as
El Hechizado, 'The Hexed'.[99] Contemporary reports describe
the baby Charles as weak and big-headed. He was unable to
speak until aged four or walk until he was eight; the frail child

was constantly carried around by a nurse. Charles suffered swelling of his feet, legs, abdomen and face, and his mouth was filled with an overlarge tongue. He showed very little interest in his surroundings – a medical condition known as abulia – and suffered fits of epilepsy. He regularly urinated blood, had intestinal problems and was afflicted with diarrhoea and vomiting.[100] The British envoy Alexander Stanhope wrote, Charles II 'has a ravenous stomach, and swallows all he eats whole, for his nether [lower] jaw stands so much out, that his two rows of teeth cannot meet. To compensate which, he has a prodigious wide throat, so that a gizzard or liver of a hen passes down whole, and his weak stomach not being able to digest it, he voids it in the same manner.'[101] In his final years, the king could barely stand, and suffered from hallucinations and convulsions.[102]

Charles II suffered such a litany of afflictions that they were almost certainly due to not just one genetic disease, but a whole suite of inherited disorders, rooted in generations of consanguineous marriages. Each individual can have up to a total of 24 great-grandparents and great-great-grandparents combined – but Charles II had only sixteen.[103] Similar figures recurred throughout his ancestry. Charles's mother was the niece of his father, and his grandmother was also his aunt. The Habsburg gene pool had become very shallow and stagnant indeed.

Charles was only three years old when his father died, and so his widowed mother was appointed as the Queen Regent to rule on his behalf, aided by personal ministers. This arrangement was restored when it became abundantly clear that even as a legal adult Charles was incapable of ruling the empire, and on his mother's death in 1696, his second wife took over the mantle.[104] The only critical task needed of the pitiful king was that most natural and innate of human functions – to reproduce. Yet although Charles was married twice, no children resulted from either union. His first wife spoke of his premature ejaculation; the second complained about his impotence.[105] It seems that Charles was congenitally unable to father any children. The generations of inbreeding and mounting recessive disorders had finally collapsed. The Spanish Habsburg dynasty was doomed before its last king had even been born.

As Charles II neared death without an heir at the end of the seventeenth century, his branch of the Habsburg dynasty faced extinction, ending its two centuries-long rule over Spain and its extensive overseas possessions. Attempts by the French and English to negotiate a partitioning of the Spanish Empire to maintain stability and a balance of power were rejected by the Habsburgs, who wanted to see their empire preserved intact. Charles II stubbornly insisted that the empire be inherited 'without allowing the least dismemberment nor diminishing of the Monarchy founded with such glory by my ancestors'.[106] Consequently, within months of the death of Charles II in 1700, the War of the Spanish Succession engulfed the continent and raged as ferociously in the colonies in the West Indies and French Canada.[107] By its conclusion in 1714, the conflict had fundamentally shifted the political landscape of Europe and the world. The French prince Philip of Anjou was crowned Philip V of Spain, retaining much of the empire; the Dutch Republic had effectively been bankrupted by the war; and Britain, having established its naval superiority, began its ascent as the dominant commercial power.*

Over the last 200 years, nation states have increasingly shifted from rule by monarchy to republics and representative democracies, with the transition of power proceeding either gradually

---

* Heeding the cautionary tale of the extinction of the Spanish Habsburgs, the Central European Habsburgs tried to ensure their own biological survival as a dynasty. Charles VI, Holy Roman Emperor, found himself the sole surviving male member of the House of Habsburg, and prudently took precautions to help ensure the smooth inheritance of power within his own family. He issued an edict in 1713 known as the Pragmatic Solution, which declared the Habsburg empire could also be inherited by a daughter (contradictory to the established Salic Law we encountered on page 59). This foresight proved invaluable, because by his death three decades later, Charles VI had fathered only three surviving daughters, the eldest of which, Maria Theresa, was able to succeed as Holy Roman Empress in 1740.

or by violent revolution. Today, of almost 200 independent states around the world, only around twenty monarchies remain, and on the whole their role is ceremonial.[108]

Rule by kingship was once intimately tied to kinship. Now, transmission of ruling power through inheritance has virtually disappeared, but the influence of family and dynasty still persists even within modern democracies. While political office is no longer hereditary, the members of political dynasties benefit from significant advantages over newcomers. Their surnames are already familiar to the electorate, they are able to draw on established networks of supporters and financial backers, and they often already have substantial familial wealth.[109]

Within India, the largest democracy in the world, political families have dominated the government since independence. In 2009, nearly a third of the elected members of parliament had relatives serving in public office at the same time or immediately preceding them.[110] Kinship is also particularly strong in the politics of Japan, South Korea, Thailand and Indonesia.[111] In the United States, two presidents have had sons become president: John Adams (1797–1801) and then John Quincy Adams (1825–1829); and then more recently George Bush (1989–1993) and George W. Bush (2001–2009). The Taft, Roosevelt and Kennedy families all also occupied the White House, as well as numerous other prominent elected government offices over a century or more. Nor are modern democracies immune to nepotism – within President Trump's administration (2017–2021), for example, both his daughter and son-in-law were appointed to prominent government posts.[112]

Family businesses have also been common through history, and dynastically inherited companies remain a significant part of today's economy. Large corporations today owned or led by multiple generations of the same family include major banks (Barings, Rothschild, Morgan), car manufacturers (Ford, Toyota, Michelin) and many other familiar firms including Heineken, Ikea, Levi Strauss and L'Oréal. The process of generational handover, with an elderly CEO refusing to step down or rivalry between siblings to take over the leadership, can be as turbulent as anything seen historically with royal succession.[113]

The origins of the human family are ancient, but their influences on our life today are as strong as they have ever been.

Another constant of human life is our vulnerability to infectious diseases, and we'll turn now to their history-defining effects.

Chapter 3

Endemic Disease

*But it seems that for our sins, or for some inscrutable judge-
ment of God, in all the entrances of this great Ethiopia that
we navigate along, He has placed a striking angel with a
flaming sword of deadly fevers, who prevents us from penetrating into the interior to the springs of this garden.*
              —João de Barros, *Decada Primeira*

A disease is by its very nature a distortion of the normal, healthy
functioning of the human body. It acts to prevent some of our
internal systems working properly, causing disorder, debilita-
tion, even death. Many diseases are caused by mutations in the
DNA coding inherited from our parents or acquired during our
lifetime, such as the copying errors that trigger cells to begin
proliferating out of control and cause cancer.

A huge number of diseases are caused by microscopically
small organisms, microbes, invading our body. Such pathogenic
microorganisms can be transmitted to humans in a variety of
ways. Some are contagious through direct contact, such as leprosy or HIV/AIDS; others are airborne and enter the throat
and lungs, such as influenza or COVID-19; others still, like
cholera, spread through drinking water contaminated with
human waste. Then there are those transmitted via an inter-
mediary or 'vector' – typically parasitic, bloodsucking insects
such as the mosquito (malaria, yellow fever, dengue), tsetse fly

(sleeping sickness), flea (bubonic plague) louse (typhus) or biting arachnids such as ticks (Lyme disease).[1]

But regardless of the mode of transmission, what pathogens share is their ability to survive in the internal environment of a body and exploit particular features of the host's biology to complete their own life cycle. Some are even able to completely evade the body's surveillance and defence mechanism, the immune system, like a con artist disguised as a contractor, complete with high-vis jacket and clipboard, walking unchallenged past the security desk. Thus there is often an intimately close relationship between the biology of a pathogen and that of its host, honed over evolutionary time so that the pathogen can reproduce and proliferate.*

The necessity of such well-tuned adaptations means that only a tiny sliver of all microbial life on Earth is able to infect and then reproduce in the human body. Of the countless millions of different species of microorganisms on the Earth,[3] only 1,128 are known to do so – around half are bacteria, a fifth are viruses, and the remaining third or so are fungi and parasitic protozoa.[4] A further 287 organisms that cause diseases in humans are not microbes but parasitic worms.[5]

And the majority of these pathogenic microbes – about 60 per cent – are zoonotic; they are transmitted to humans from animals. Thus at the source of the great pestilences and plagues that have stalked humanity since the dawn of civilisation are the wild beasts we domesticated as livestock and began living in

---

* The SARS-CoV-2 virus that caused the COVID-19 global pandemic is a pertinent example. The coronavirus resembles a ball with protruding suckers called spike proteins, which are perfectly shaped to dock onto a specific molecule found on the outer surface of cells lining the respiratory tract. This causes the cell to bring the virus inside, where it hacks the messages of the cell's own construction orders and subverts the genetic machinery to instead replicate hordes of new viral particles that then burst out into the lungs. The different variants of the COVID-19 virus that have emerged tend to have mutations in that spike protein, as the virus has evolved to invade cells more effectively and enhance its transmission to more hosts.[2]

close proximity with. Dispersed bands of hunter-gatherers are generally healthy, save for parasites, and as long as they survive childhood many can expect to live to over 60 years old.[6] Most of the diseases that affect them are not infections caught from one another, but 'wear and tear' afflictions such as arthritis and rheumatism.[7] There are some ancient infectious diseases that predate the development of agriculture, such as malaria and leprosy,[8] which are able to sustain themselves in small, dispersed populations of humans or survive in animal hosts if susceptible humans are scarce. But once humans started settling down in ever-denser concentrations in agricultural villages and then cities, we created the perfect breeding conditions for infectious diseases to jump not only the species barrier but from person to person. History saw a surge of 'crowd diseases'.

When a pathogen first breaks into a new population, in which individuals' immune systems are unprepared for the attack, it can spread very quickly as an epidemic with a staggering initial death rate. But many diseases wane in severity over time, as the population acquires resistance or the pathogen mutates. They may circulate among the population as a constant presence, perhaps flaring up occasionally in outbreaks – a simmering, background burden rather than a raging firestorm. Some diseases can disappear completely. English sweating sickness, for example, appeared in London in 1485, probably brought by Flemish mercenary troops who had helped Henry VII seize the throne from Richard III at the end of the Wars of the Roses. It raged across Britain, seeming to preferentially strike rural middle-aged men of the more affluent social classes during summer,[9] with an extremely sudden onset and high mortality – death occurred typically within just hours. The mysterious contagion recurred in a series of outbreaks over seventy years, before vanishing entirely in the mid-sixteenth century, never to return.[10]

In the past, a widespread disease outbreak would be termed a plague or pestilence. Nowadays, the more technical term is 'epidemic' – derived from the Greek *epi*, 'among', and *demos*, 'the people' – or, in its most extreme case, 'pandemic', a disease that has spread to many people across a very large area. 'Endemic', denoting a disease found in a particular area, derives from the Greek for 'within the people'. Both endemic and epidemic diseases have had far-reaching impacts on whole societies and civilisations, so I decided to dedicate two chapters to exploring their effects on human history. The difference between the two is not a matter of taxonomy, however: the same pathogen can swirl around a partially resistant population with only minimal impact and then break into a new population to trigger a devastatingly lethal plague. One region's endemic disease is another population's epidemic. But I believe the different effects that the two patterns of disease have had on human history still make for a sensible distinction.

We'll look first at how human susceptibility to endemic diseases has played a significant role in world history.

THE DARIEN SCHEME

In the late 1600s, Scotland was struggling. Its economy, based almost entirely on agriculture, had been suffering years of poor harvests and famine. Despite the fact that England and Scotland had been ruled by the same monarch since 1603 – when Queen Elizabeth I died without an heir and her distant cousin James VI of Scotland succeeded to the English crown – Scotland remained fiercely independent of its powerful southern neighbour. The English, in turn, imposed oppressive economic restrictions on the Scots, including protectionist trade bans with France and the North American colonies. In order to place itself on a more secure economic footing, and avoid being forced into an unfavourable union with England, the Scottish people began to look further afield. England had grown rich on foreign trade, and the Scots yearned for a piece of this mercantile pie that could deliver such handsome returns. Previous Scottish settlement attempts – in Nova Scotia, East New Jersey and South Carolina – had not been successful, but the solution to turning around the country's economic and political fortunes was still considered to lie in the establishment of its own overseas colony to capitalise on maritime trading routes.

An ambitious plan was hatched, masterminded by the Scottish-born financier (and one of the founders of the Bank of England) William Paterson. The Scots would establish a colonial settlement on the Isthmus of Panama, the narrow thread of land linking the North and South American continents. With a port and trading post positioned here, they reasoned, they would be able to participate in the vigorous commercial network strung between the islands of the Caribbean, and across the Atlantic to Africa. But Paterson also had further, even more ambitious, plans for the colonial endeavour.

At the time, ships from Europe or the Atlantic coast of North America heading west to China and the Spice Islands had to sail all the way down the coast of South America, around Cape Horn and back up again to cross the Pacific – a huge continental diversion. So why not build a road across the thin, 80-kilometre-wide sinew of the Panama isthmus that could link shipping on both sides – a shortcut to convey goods between the two greatest oceans of the planet? It would slash the time needed to reach the Orient, and therefore the expenses involved, by over half.[11] Ultimately, Paterson envisaged digging an artificial water route for shipping to pass directly between the oceans – a Panama canal. And with the Scots controlling this gateway between the Atlantic and Pacific they could secure a sizeable income from the tariffs charged on the passage of cargo. The location that Paterson had carefully chosen for this trading colony was on a small peninsula, with a well-protected bay for shipping, in the Darien region of the Panama isthmus. So this daring plan to transform Scotland's destiny became known as the Darien Scheme.

The colonial endeavour was orchestrated through the newly established Company of Scotland Trading to Africa and the Indies, which, it was hoped, would grow to rival England's East India Company. The company quickly secured around 1,400

Scottish investors, drawn from right across the social spectrum, ranging from MPs to farmers. It has been estimated that between a quarter and a half of all liquid capital in Scotland at the time was poured into this bold, enterprising venture.

In July 1698, five ships set sail from Edinburgh loaded with 1,200 colonists and the hopes of a nation. The cargo holds were full of the materials, tools and equipment needed to establish a new colony from scratch, along with provisions for the voyage and first months of arrival. The settlers had been carefully chosen for the different skills needed at their destination. They arrived there at the end of October and founded the colony of New Caledonia, with New Edinburgh as its capital and defensive fortifications built on the sheltered peninsula.

Word was sent back to Scotland that the colony had been successfully established and was flourishing, enjoying peaceful relations with the indigenous population. Yet these first letters were whitewashing the truth in order to encourage further ships and supplies to be dispatched. The reality was that New Caledonia was already in dire straits.

The settlers soon discovered that the terrain inland was far too difficult to permit an overland coast-to-coast passage – and there certainly was no prospect of being able to dig a canal to link the oceans. Nonetheless, establishing a profitable entrepôt embedded in the bustling trade routes in the region was still seen as a viable possibility. But far more serious a problem was that almost immediately the colonists began succumbing to the diseases rife in the region. During his fourth and final exploratory voyage to the Americas in 1502, Columbus and his crew had been so tormented by insect bites along a stretch of the isthmus that they dubbed the region the Mosquito Coast.[12] The Scots had arrived in peak mosquito season and were soon ravaged by mosquito-borne diseases – malaria and yellow fever.

Malaria is the disease that has probably been with humans the longest and killed the most. The infection is characterised by a fever that begins with intense chills and uncontrollable shivering, leading to a spike in temperature and then profuse sweating and fatigue, with the pattern repeating every few days. It is caused by a single-celled parasite – a kind of plasmodium – that is transferred from the blood of one individual to the next by the bite of a mosquito. The fact that malaria is spread by a flying vector means that, unlike crowd diseases, it doesn't require dense populations of people to sustain itself, and so it is thought to predate the emergence of agriculture. In fact, malaria may be a truly ancient disease – the parasite that infects us probably evolved from those afflicting our closest evolutionary cousins, the great apes, in the tropical rainforests of Africa.[13] Malaria had long been endemic over much of sub-Saharan Africa but was brought to the Americas only with European contact, probably aboard early slave ships from Africa.[14]

Yellow fever is a viral disease and also originated in Africa, jumping into humans from primates some 1,500 years ago.[15] The first definitively classified epidemic in the Americas occurred in Guadeloupe in 1647,[16] although the virus first arrived in the previous century aboard slave ships from Africa, before spreading widely across the Caribbean and two continents, as far north as Quebec.[17] The early symptoms of the disease include fever, muscle pains and headaches, and in more severe cases this progresses to liver and kidney damage with a high mortality rate. Its name refers to the jaundice resulting from liver failure, and those who don't die of haemorrhagic fever – in Spanish it is known as *vomito negro* on account of the tar-coloured vomit caused by internal bleeding[18] – recover completely with life-long immunity.[19]

The New Caledonian colonists were thus hit with a double whammy of lethal diseases. Within six months they had lost almost half of their initial number, with as many as a dozen dying every day in the small settlement.[20] The expedition had met with ruin, and the surviving settlers returned to their ships and abandoned the colony in July 1699, leaving behind those who were too weak to move. But the fleeing colonists continued to die in droves at sea: only 300 out of the original 1,200 settlers survived the ordeal.

However, word of New Caledonia's abandonment had not

reached home before a second wave of resupply ships was dispatched bearing extra provisions and 300 more settlers. When they reached Darien, they found a ghost town of empty huts and overgrown farm fields. They promptly turned around and sailed home. But still, news of the failure did not arrive in Scotland soon enough to prevent another large fleet of colony ships, carrying more than 1,200 further settlers,[21] setting sail for Darien. This group of colonists stayed, but the second attempt at making New Caledonia a success fared no better than the first, and within months around a hundred were dying every week of malaria and yellow fever, with the colony now also being harassed by Spanish raids. In April 1700, the survivors surrendered to Spain. Of this second wave of colonists, fewer than a hundred made it back home. New Caledonia was abandoned for good, and the Scottish dream of an American colony and overseas trading riches collapsed with it.

The Darien Scheme had abjectly failed. Perhaps as many as 80 per cent of the 2,500 Scottish settlers who had sailed to New Caledonia were killed by malaria and yellow fever, their dire situation compounded by isolation from the English colonies and outright hostility from the Spanish.* Had the Scottish colony been able to successfully link the Atlantic and Pacific Oceans, or at least maintain their strategically placed entrepôt to challenge the regional trade dominance of England and Spain, it may have changed the course of history. In the event, the dream of digging an artificial waterway to link the Atlantic and Pacific Oceans was not realised for another two centuries after the failure of New Caledonia.†

* The Scots were not disproportionately unlucky with the ravages of tropical diseases. Spain's colonies in the region were also hit hard – between 1510 and 1540, an estimated 40,000 Spanish settlers died on the Mosquito Coast, mostly of the two tropical diseases[22] – but the Spanish were able to sustain these losses with replacements.

† In fact, history has witnessed several proposals for a Panama canal. The Spanish contemplated the scheme in the 1530s to speed up travel between Malaga and Peru and outmanoeuvre the Portuguese, and again in the late

With the loss of the colony, the huge amount of investment that had been raised in Scotland to support the venture had evaporated. Indeed, the failure of the Darien Scheme brought Scotland to the brink of financial ruin and was a decisive factor in forcing Scotland into a political union with England. For a century after the Union of the Crowns in 1603, Scotland had remained an independent kingdom with its own parliament. This autonomy was now threatened by the dire financial straits Scotland found itself in. England promised assistance by reimbursing the Company of Scotland shareholders and ending the economic restrictions on trade.[25] This was an irresistibly tempting offer for the Scottish elites – the aristocracy and mercantile class that had lost heavily in the economic fallout of the failed Darien venture – who felt their best option was to tie their future to the growing English trade empire and its international might. Six years after the loss of New Caledonia, the Scottish parliament was left with no choice but to consent to the union with England.

Scotland had surrendered its sovereignty, and Great Britain was born, as a result of the mosquito-borne diseases in a remote part of Panama.[26]

1780s; the British considered a plan in 1843. The French, buoyed by their recent success with the Suez Canal, began a concerted excavation effort with steam shovels to excavate across the Panama isthmus in 1881. But the effort failed before the decade was out with the deaths of around 22,000 workers, overwhelmingly from malaria, yellow fever and other tropical diseases.[23] The French hadn't encountered insurmountable geographical impediments or engineering problems: their canal project was abandoned for the very same reasons that New Caledonia failed – the punishing disease environment of this swampy region. The Panama Canal was finally completed by the United States between 1904 and 1914 – passing just 200 kilometres from the ruins of New Edinburgh. This only became possible once the mode of transmission of yellow fever and malaria had finally become understood, and aggressive mosquito control measures – including drainage and kerosene spraying – were enacted along the length of the excavations.[24]

REVOLUTIONARY FEVER

Europeans had initially benefitted from an enormous epidemiological advantage in their conquests of the New World from the early sixteenth century: as we'll see in the next chapter, they brought with them pathogens that wiped out indigenous populations. But over the following centuries, the tide turned.

Endemic diseases are often thought of as a curse. An inescapable and ever-present sickness stalks the land like an evil spirit, exacting its toll on the health of the population and in particular on young children. But with many endemic diseases, those who survive to adulthood are bestowed with lifelong immunity, or at least greater levels of resistance. Endemic disease can therefore protect a seasoned native population against foreign invaders. In their native disease environment, defenders benefit from a home ground advantage compared to the susceptible intruders. European armies arriving to quell uprisings within their colonial lands in the Americas found they were at a distinct disadvantage, succumbing to the endemic diseases at far higher rates than their opponents, especially in tropical climes where many of the cash crop plantation systems had been established. And the ramifications of this disease biology has had profound consequences for how history has played out.

The American Revolution grew out of years of mounting tension and discontent among British colonies along the Atlantic coast over imperial Britain's squeezing of their autonomy and imposition of taxes without representation in parliament. In late 1774, the Thirteen Colonies formed the Continental Congress to coordinate their resistance to British rule, which erupted into open conflict the following spring. The colonies had united in opposition and soon declared their independence; now they just needed to win the war.

At the start of the American War of Independence, the British Army was one of the most highly trained, best-equipped fighting forces in the world. Many of the red coat soldiers were battle-hardened from action around the world against the French and Spanish during the Seven Years War the previous decade. Though Britain was still cash-strapped after this global struggle for imperial supremacy, it was on a far stronger economic footing than the Thirteen Colonies. And British command of the Atlantic Ocean by the Royal Navy allowed it to strike right along the North American seaboard and blockade the small American fleet in port, choking off imports of food and war matériel. In contrast, the revolutionaries started with only militias of self-trained civilians, before the Continental Army was formed a couple of months into the war.

The British were able to push home their military advantage in the early stages of the conflict. They quickly captured the principal colonial ports of Boston and New York but were unable to land a decisive victory against the Continental Army to end the rebellion. The revolutionaries manoeuvred around the countryside, deftly avoiding being drawn into pitched battle and annihilated before they could secure more American support or the involvement of other foreign powers. The colonists didn't win their first big victory until two and a half years into the war, at the Battle of Saratoga in New York state in October 1777. This demonstrated to the world that the Americans had a fighting chance, and the French and then Spanish now entered the conflict on their side. French warships were able to crack open the Royal Navy blockades; the Spanish supplied arms and provisions through the port of New Orleans; and the arrival of professional French soldiers in the second half of the war helped tip the balance in the colonists' favour.

When stalemate had been reached in the north, the British decided to try a new tack, launching a fresh strategy at the end of 1778 in the south. The hope was to recruit large numbers of loyalists in the most recently founded colonies of Georgia and the Carolinas, secure the profitable plantations, and crush the rebellion once and for all.

After initial successes, the British commander-in-chief General Henry Clinton left General Charles Cornwallis in command of the southern regiments of 9,000 soldiers and returned to defend New York against an anticipated counter-attack. Despite these victories, the tide of the war was already turning against the British. The southern strategy had committed a large proportion of their forces to the mosquito-ridden subtropical regions, exposing them to the onslaught of malaria and yellow fever – enemies that the British were ill-equipped for.[27]

Lifetime exposure to diseases like malaria and yellow fever bestows a degree of immunity, but medicine can also treat or prevent infections. The bark of the cinchona tree was known to be effective against malaria, but it was in short supply.* At the outbreak of the war, George Washington, commander of the Continental Army, had urged the Continental Congress to buy up as much as possible.[31] The British, on the other hand, were acutely short of this vital preventative medicine. At this time, the only source of cinchona was high in the Spanish-controlled Peruvian Andes, and the Spanish had completely cut the British off from this supply in 1778, shortly before they joined the war alongside the French in support of the American Revolution. In addition, the British had committed much of their quinine reserves to their troops maintaining order in India or fighting imperial scuffles in the Caribbean. Throughout the southern campaign, therefore, Cornwallis's officers and men were plagued by malaria.[32] The soldiers of the Continental Army were not immune to malaria, but having previously lived with local

---

* The cinchona or 'fever tree', which grew natively only in small, isolated pockets of the Andean mountain range in South America, contains the active compound quinine in its bark.[28] The indigenous Quechua people took the ground bark as a herbal remedy, although for treating the febrile shakes of severe chills rather than malaria specifically.[29] Jesuit missionaries brought cinchona bark back to Europe in the mid-seventeenth century, where it was used to treat, for example, the Pope in Rome suffering from malaria (mosquitoes flourished in the Pontine Marshes formed in the alluvial plains along the coast south-east of the city). Powdered cinchona bark was then used throughout European colonies beleaguered by malaria, but the supply remained limited. We now understand that quinine is able to prevent or treat malaria because it is toxic to the single-celled plasmodium that causes the disease.[30]

strains of the disease they were less severely affected – they had the home ground advantage of an endemic disease.

Cornwallis kept his army on the move across the Carolinas, trying to find areas that might offer a reprieve from the 'miasmic diseases', especially during the peak mosquito season from late June to mid-October.[33] By the time Cornwallis had won the Battle of Camden in mid-August 1780, many of his soldiers were crippled by 'fevers and agues' and too weak for service.[34] Throughout the early months of 1781, Cornwallis chased the American revolutionary forces around the Carolinas. All the while, the Americans constantly harried the British southern army with short skirmishes before withdrawing, denying the soldiers any rest and driving them to exhaustion. And while they conducted their hit-and-run, guerrilla-style tactics, the local mosquitoes kept up their onslaught against the unseasoned British troops. By April, the number of Cornwallis's men fit for service had almost halved.[35]

When the main force of the combined French-American army marched into Virginia in late summer 1781, Clinton ordered Cornwallis to withdraw his southern army into Yorktown, on the coast of Chesapeake Bay, to construct a defensive position and await evacuation by the Royal Navy. Clinton still believed that Washington intended to attack New York and wanted to keep the southern army within reach of the navy if a rapid redeployment north became necessary. Cornwallis repeatedly questioned the wisdom of these orders to 'hold a sickly defensive post in this bay', surrounded by marshy estuaries and with growing numbers of his men falling prey to the fevers.[36]

The Franco-American army laid siege to Yorktown, and when the Royal Navy fleet sent from New York arrived in early September 1781, it was repelled by the French fleet guarding the mouth of Chesapeake Bay. Cornwallis was stuck with the worst of both worlds: his army pinned down on the coast during peak mosquito season and the Royal Navy still not able to reach him.[37] With over a third of his remaining army too sick to fight,[38] Cornwallis laid down arms on 19 October 1781. His surrender ended the war: the United States had won its independence.

The American revolutionaries would not have prevailed without the intervention of the French and Spanish in their favour. Their supply of weapons, provisions and reinforcements, as well as their fleets challenging the Royal Navy and breaking the blockades, were all key to the success of the revolutionary struggle. But also hugely influential was the weakening of British forces by endemic diseases during the southern campaign – their susceptibility to malaria only compounded by the shortage of quinine. The American soldiers, on the other hand, benefitted from a home ground advantage bestowed by their lifelong sea-soning to local endemic diseases.[39]

RESISTANCE

It wasn't just the Thirteen Colonies that owed their success in the fight for independence in part to endemic diseases. Shortly afterwards, a similar scenario played out when the slaves in the French colony of Saint-Domingue, on the Caribbean island of Hispaniola, rose up against their masters.[40]

Hispaniola is the second largest island of the Caribbean after Cuba and was the site of the first European settlement in the Americas, established on Columbus's world-changing voyage. But the Spanish largely lost interest after they realised that the island had no gold or silver to offer. The French established themselves before formally gaining the western third of the island in the peace treaty that concluded the Nine Years War in 1697. The French named their colony Saint-Domingue (which is now Haiti), and by 1775 it had become the most lucrative in the world.[41] Saint-Domingue's 8,000 plantations grew half of the global supply of coffee; it was also the world's top producer of sugar and a leading exporter of cotton, tobacco, cocoa and indigo dye. This single colony accounted for over a third of France's entire trade and produced a greater economic output than the thirteen British mainland colonies combined.[42]

For all that, Saint-Domingue depended on the transatlantic

slave trade. The work of cultivating cash crops in the tropical heat was gruelling, and the staggeringly high death rate among the slaves demanded constant replenishment. By the late eighteenth century, around 30,000 new slaves arrived every year, a constant influx needed to sustain the total numbers at around half a million. Slaves made up some 90 per cent of the colony's inhabitants.

In August 1791, a group of them rose up against the brutal oppression of the plantation owners and violent revolt rapidly spread across the colony. Within weeks, their number had swollen to 100,000, and the following year the rebels controlled a third of the colony.

The British were unsettled by the slave revolt: if it were allowed to succeed, working slaves on other colonies would be encouraged to rise up too, leading to a domino effect of rebellions across the Caribbean. Being at war with France, the British also recognised an opportunity to secure the exceedingly lucrative French colony for themselves. But the conflict proved disastrous. British troops arriving on the island with no previous exposure to or resistance against tropical disease also succumbed in droves: around 65 per cent of the 23,000 British soldiers sent to Saint-Domingue died of yellow fever or malaria.[43]

After military intervention by the British, as well as the Spanish, a former slave, Toussaint L'Ouverture, had emerged as the most prominent leader of the Haitian Revolution and consolidated control of the colony, issuing a constitution and calling for Saint-Domingue to be an independent black state. He had earned his second name – meaning 'The Opening' – with his ability to find the ways through enemy ranks; both friends and enemies also referred to him as Black Spartacus or Black Napoleon.

In 1801, Napoleon himself dispatched General Leclerc, his son-in-law, with over 25,000 soldiers to quash the slave rebellion and regain control of the lucrative colony.[44] At first, the highly trained and well-equipped French troops were successful in their engagements against the slaves and L'Ouverture was captured. But like the Americans two decades earlier, the rebel army harried the French from the inland hills with hit-and-run tactics of guerrilla warfare, confining them to the low-lying coastal regions.[45] And these freedom fighters too were aided by the squadrons of mosquitoes on the island and the endemic diseases they transmitted – malaria and yellow fever. What's more, another key biological difference between the African rebels and the European soldiers came into play.

As we have seen, the human body develops resistance to the malarial parasite after repeated infection – provided the individual survives the experience – and in malarial areas, surviving children develop a significant acquired immunity, known as seasoning, by age five.[46] To misappropriate Friedrich Nietzsche: 'That which does not kill me, makes me stronger.'[47] But malaria has imposed such a burden upon exposed populations that different groups of humans have evolved with genetic mutations that confer innate resistance against the disease. These defences mostly affect the red blood cells, where the malarial parasite grows.[48] Unsurprisingly perhaps, they're mostly found in Africa, where malaria is endemic and humanity has spent the great majority of its evolutionary history. The most important of these mutations that provide shielding against malaria is that which causes sickle cell anaemia.* The sickle cell trait is caused

---

* Other genetic shields against malaria include the Duffy negative blood group, thalassemia and a deficiency of an enzyme called G6PD. The Duffy antigen is a receptor molecule on the outside of red blood cells that the malaria parasite uses as a gateway to invade the cell – so an absence of this antigen, knocked out by a mutation, blocks the parasite's entrance and prevents it completing its life cycle. Duffy negativity is found in around 97 per cent of West African and western Central African populations, giving them resistance to certain forms of malaria (although it has been shown to increase susceptibility to a far more recent disease to hit Africa, HIV.)[49] Thalassemia is a disorder that reduces the production of haemoglobin in the blood and is particularly prevalent among populations in the Middle East, North Africa and Southern Europe. G6PDD is the deficiency of the G6PD enzyme that helps red blood cells scrub out destructive oxidants, and it is particularly

by a mutation in the gene that produces haemoglobin, the vital molecule inside red blood cells that makes them red and carries oxygen around the body. Everyone has two copies of this gene, one received from each parent. Individuals who inherit one normal version of the gene and one allele with the sickle cell mutation – who are said to be 'heterozygous' – are carriers of the disease.

Normal red blood cells are shaped like a thick disc compressed at the middle. But if they encounter conditions of low oxygen, the haemoglobin molecules produced by the mutated gene clump together and deform some red blood cells into a sickle-shape. Sickle cells can then get jammed in narrow vessels and block blood flow. Carriers of the sickle cell trait usually don't experience many adverse effects – except in severe cases of oxygen starvation, such as strenuous exercise or, in modern times, unpressurised aeroplane flights.*

However, having one copy of the sickle cell allele is exceedingly effective at protecting the carrier against severe cases of malaria – it either hinders the parasite's growth within the red blood cells, or makes it more likely that the affected cells are destroyed by the immune system.[54]

The problem is that individuals who receive two copies of

---

common around the Mediterranean and Middle East. Carriers do not experience any particular negative effects, unless a trigger causes their red blood cells to suddenly break down. Compounds found in certain foods can serve as such a trigger – including fava or broad beans, which may be why the Greek philosopher Pythagoras warned against the dangers of eating this particular vegetable in the sixth century BC.[50] Exactly how thalassemia and G6PDD protect against malaria is not entirely clear, but they are thought to hinder the malaria parasite growing or infecting more red blood cells, or to result in infected cells being removed more quickly by the immune system.[51]

* Sickle cell trait therefore poses a risk of bouts of extreme pain, or even sudden death, to those of African descent during military training or playing sports.[52] The trait has also been used to exonerate officers involved in the deaths of black detainees while in police custody – the argument being that restraining techniques that restrict breathing inadvertently caused sickle cell crisis.[53]

the mutated gene from their parents – and are therefore homozygous – suffer dire consequences. The distortion of their red blood cells is much more common and they are stricken with sickle cell disease, which causes anaemia and blocks blood flow to the organs. Without modern medicine, those who are homozygous do not survive to adulthood. The sickle cell mutation is therefore a double-edged sword: it protects heterozygous carriers from malaria, but homozygous individuals suffer an affliction just as bad, if not worse, and die young.

In malarial sub-Saharan Africa, then, natural selection has been pulled in opposite directions. The outcome of this evolutionary tug-of-war between sickle cell disease and malaria is an equilibrium in the population: some 20–30 per cent of the population of malaria-infested Africa are carriers of the sickle cell variant of haemoglobin.[55] The fact that the mutations to the genetic code causing sickle cell – a horrifying disease in its own right – would be so powerfully selected reveals a desperate Darwinian battle with malaria.[56] The existence of sickle cell disease in the world today, affecting some 300,000 children born every year[57] – mostly to parents of African descent – is the cost of evolution struggling to protect populations against the most destructive disease in human history.*

---

* There are other interesting examples of genetic or cellular differences responding to diseases that have plagued human populations for millennia. People with blood type O are more susceptible to severe infections of cholera; so perhaps unsurprisingly, Bangladesh, where cholera has been endemic in the river water for thousands of years, has the lowest prevalence of this blood type in the world.[58] In other cases, genetic differences that provided resistance against past epidemics are useful in protecting against newly emerged diseases. For example, a mutation in a protein sat on the outside of white blood cells, which receives the signalling molecules used by the immune system, provides resistance against HIV/AIDS to the few per cent of Europeans who carry it but is believed to have originally been strongly selected for because it protected against bubonic plague or smallpox.[59]

The genetics behind cystic fibrosis are very similar to that of sickle cell disease. There is a gene that codes for a transporter molecule that regulates the movement of salt and water in and out of cells. Heterozygous carriers with

But let's return to the Haitian Revolution and the slaves' bio-logical advantage over their European oppressors. As a result of their long evolutionary history in Africa, many of the slaves on Saint-Domingue would have carried an innate, genetic shield against malaria. What's more, as a result of the particularly high influx of new slaves imported to the colony each year, most had also been born in Africa and so would have been exposed to malaria and yellow fever from a very young age. In short, they were already well-seasoned to malaria, and likely possessed life-long, acquired immunity to yellow fever after having survived infection as a child.

The French troops, on the other hand, were protected by no such biological shields and suffered far more severely from both diseases. Before long, over a third of the French soldiers were sick. It seems that many of those that survived the yellow fever were then killed by malaria. General Leclerc himself died of yel-low fever. Reinforcements were sent but they too succumbed to the mosquito-borne diseases in huge numbers.[61] In total, around 50,000 French troops of the expeditionary force died on Saint-Domingue, most of them of malaria and yellow fever. Only a few thousand soldiers survived to be evacuated in 1803 when Napoleon abandoned the effort to reclaim the colony.[62]

The freed slaves, with aerial support from the common mosquito, had defeated not one but two of the most highly trained and best-equipped armies in the world.[63]

---

only one defective allele are mostly healthy, whereas babies who inherited two defective copies (and so have no normal, working copy of the gene) suf-fer acutely from cystic fibrosis, with mucus building up in their lungs and intestines. Without modern medical intervention they would rarely see their first birthday. Such a lethal mutation should be quickly purged from the population, but curiously, around 2 per cent of Europeans are carriers – it seems that some selective pressure in our recent history favoured the prolif-eration of this mutation through the generations. The best explanation is that, historically, carriers of the cystic fibrosis mutation received a degree of protection against typhoid fever or tuberculosis, both caused by bacteria attacking the digestive system or lungs.[60]

Saint-Domingue declared independence in 1804, the free nation renaming itself Haiti.[64] But though now free of the shackles of slavery and imperial control, Haiti had become an outcast. The European imperial powers did all they could to diplomatically isolate and economically choke the development of the newly independent state. They imposed trade embargos on the island's exports, and the French deployed gunboat diplomacy to exact compensation for their loss of revenue – the iniquity of freed slaves being forced to pay reparations to their former masters. The 'debt' was not repaid until the 1950s. Once the most prosperous colony in the world, Haiti is among the poorest nations on Earth today.

Yet the successful slave revolt bolstered the global abolitionist movement, and by the time of Haitian independence, all the northern US states had prohibited slavery; Britain banned the transatlantic slave trade three years later.[65] But the effects of the Haitian slave revolt, aided by the endemic diseases of yellow fever and malaria, had other, far-reaching repercussions.

The hugely profitable Caribbean colony of Saint-Domingue had not only provided a crucial income to France's coffers, it had also been regarded by Napoleon as a critical staging post for projecting military force into North America. The entire Mississippi river valley had been claimed in the late seventeenth century by French explorers who named it after their king, Louis XIV, with New Orleans as the capital of the Louisiana colony. Now the loss of Saint-Domingue both as a source of trade revenue and as a strategic naval base forced Napoleon to abandon any imperial designs on North America and instead focus on the European theatre of war. He looked to offload not only the port of New Orleans but also the whole colonial territory of Louisiana to raise cash for his European wars.[66]

The United States was reliant on trade along the Mississippi and through New Orleans, and over a third of their exports were flowing through the Gulf of Mexico.[67] When US president Thomas Jefferson therefore approached France to buy New Orleans for $10 million, he was surprised to be offered the entire colonial territory for just $15 million (equivalent to $366 million today).[68] The Louisiana purchase was completed in May 1803,[69] with the US acquiring a huge tract of land stretching from the Mississippi to the Rockies and from the Gulf of Mexico as far as Canada. The United States doubled in size with the stroke of a pen, and for only $170 per square kilometre in today's money. So effects of endemic disease in Saint-Domingue, and the biological resistance of the black revolutionaries, played no small part in the course of world history.

DISEASE AND DEVELOPMENT

The presence of endemic diseases in the tropical and subtropical regions that Europeans colonised from the sixteenth century hindered any large-scale settlements. The very high mortality rates among any Europeans living there led the colonial powers to operate extractive strategies that were designed with the primary objective of producing and then exporting profitable commodities like sugar, coffee and tobacco as quickly as possible. They were not interested in long-term development or creating an infrastructure beyond the bare minimum needed to plunder this natural wealth. The workforce was formed of slaves, and coercive control was imposed with an iron fist to maintain the outflow of profits. The few Europeans present in these colonies mainly served as administrators or soldiers to supervise the extractive operation and suppress resistance. Long after independence from colonial rule, therefore, such areas continued to be hampered by poor infrastructure and the lack of well-established legal and governmental systems that protect against expropriation of property and prevent abuses of state power. Many of the modern states in Africa, Asia and Latin America that have endured a history as extractive colonies are today among those with the lowest levels of development or economic stability.

On the other hand, colonies established in more temperate climates without high levels of lethal endemic diseases allowed the survival of European settlers. Such colonies were able to attract significant numbers of émigrés trying to build new lives for themselves and their families. They also brought with them their expectations of just government and protection through enforcement of the rule of law. Property laws protected the rights of individuals to be able to earn a livelihood by farming their own land, through claims over mineral rights and mining or through trade. They resisted the overreach of governmental power and built representative democracies for a more equitable society, which replicated key administrative, legislative, judicial and educational institutions of the colonial homeland. In short, settlers strove to recreate little European bubbles in remote colonies. The colonies where European settlers could survive and thrive established long-lasting institutions – though only for the benefit of the colonists and their descendants. These in turn created more stable social conditions that fostered ongoing investment, development of infrastructure and economic growth even after the settler colonies achieved independence from their imperial founders. The colonies that would become the United States, Canada, New Zealand and Australia all benefitted from these drivers for long-term development, born of the fact that European settlers were not severely ravaged by endemic diseases in these regions.

So the forms of colonies that developed in different regions of the world – extractive and settler colonies – were largely dictated by biological factors: the susceptibility of Europeans to the endemic disease environment. These initial conditions established contrasting patterns of development that either helped or hindered long-term economic trajectories, and still persist now, long after the colonies achieved independence as their own nation states. The economic disparity between many nations today has its roots in where the European colonists had a self-interest in establishing strong institutions. The GDP of these nations today is strongly correlated with the historical mortality rate of European settlers.[70]

THE SCRAMBLE FOR AFRICA

The first civilisations to bear the brunt of the age of European conquest and colonisation were those of Central and South America at the dawn of the sixteenth century. From the seventeenth to the nineteenth century, Europeans settled the North American continent, exterminating its indigenous peoples in the process, fulfilling what they considered to be their manifest destiny. Western imperial interests turned to India in the eighteenth and nineteenth centuries, and targeted China in the mid-nineteenth, but did not attempt to penetrate into the deep interior of the vast continent of Africa until the late nineteenth century.

As we have seen, prevalent diseases shaped the colonisation – and history – of the Americas. Yet when European explorers first made contact with the New World, they encountered an effectively benign disease environment. We need to remember that malaria and yellow fever, which had such devastating effects on colonists, were not originally present in the Americas, but arrived with European contact and the slave ships from Africa. For historical reasons we'll turn to in the next chapter, native American societies had few crowd diseases of their own, and the first settlers faced little hindrance from endemic disease. On the other hand, the infectious diseases that the Europeans unwittingly brought with them wrought apocalyptic devastation upon the native populations. Whole civilisations collapsed beneath the lethal onslaught of Old World diseases, bringing far more death and destruction than the weapons of the invading troops. The settlers following behind encountered largely emptied territory.

But the European powers benefitted from no such epidemiological advantage in their engagements with India, China and other societies across Asia. Trade networks operating across the breadth of Eurasia over the millennia had ensured a thorough distribution and mixing of infectious diseases, creating a single, common disease pool. European and Asian populations had acquired resistance against the same diseases. The imperial clash with India and China relied on the superior weaponry and the formidable armies and navies of European powers.

In Africa, the power balance was reversed. Having endured lethal tropical diseases for millennia, African populations exhibited much greater genetic resistance, such as the sickle cell trait, and lifetime seasoning. So the most lethal African diseases to outsiders, malaria and yellow fever, had relatively low mortality rates among African adults. And unlike many of the crowd diseases endemic across Eurasia – such as smallpox, influenza and measles – that can be directly transmitted from person to person, many tropical diseases are spread via an insect vector. These key intermediaries cannot survive in colder environments and so the diseases are confined within particular climate zones.[71] Europeans arriving in Africa possessed no genetic resistance or seasoning to malaria, nor immunity to yellow fever, and quickly died in large numbers. Indeed, the mortality rate from yellow fever in adults who have not been previously exposed to the disease could be as high as 90 per cent.[72] Europeans may have had a stark military advantage over the native polities, but they consistently lost against the indigenous microbes. At least to start with, endemic diseases seemed to level the playing field between the native peoples and would-be invaders, serving as a very effective biological deterrent against incursion.

Yet this didn't prevent European powers from pursuing some of the worst forms of extraction from Africa. A modern map still bears the marks of what European powers considered of value: Pepper Coast, Ivory Coast, Gold Coast, Slave Coast. But the European presence in Africa was limited to a small number of troops and traders in the coastal forts – known as factories – negotiating deals with the local chiefs and overseeing the exploitation of resources from further inland, including the most abhorrent form of extractive practice, the seizing of forced human labour.[73]

Thus Africa's endemic disease environment served as an effective firewall, largely keeping out European intruders and preventing the widespread colonisation that had occurred in the Americas and Oceania. European colonisers were able to cling to no more than tiny toeholds along the coastline, and even within the factories the European death rate exceeded 50 per cent every year.[74] It was the Portuguese who first established trading posts on the West African coastline in the late fifteenth century, but for the next four centuries or so Africa remained the 'Dark Continent', its interior unknown to European eyes. For Europeans, Africa was a death sentence:[75] the British referred to the continent as the 'White Man's Grave'.[76] As late as 1870, very few Europeans dared travel more than a day or two inland from the coast.[77] Only in regions where the disease environment was unusually favourable, such as in the very southern tip of the continent around Cape Town, were Europeans able to establish a permanent presence.[78]

All this started to change from the second half of the nineteenth century. Medical science came to understand the basis of different infectious diseases – by isolating and identifying the microbe that causes each one – and so was better equipped to treat or prevent them. Labs developed new vaccines and, later, antibiotics. Production was increased of natural plant products known to combat diseases, and chemists learned not only how to extract and purify their active compounds, but then to artificially synthesise them in bulk. The ancient relationship between humanity and diseases was being fundamentally transformed, improving the lives of countless millions. But the colonial powers also used these new medical capabilities to extend their reach across the world.

From the early nineteenth century, quinine was used to aid British imperial control over malarial regions of India. In the 1860s, the British smuggled cinchona plants and seeds out of South America to establish plantations in British India and Sri

Lanka for their own supply. As powdered quinine tastes extremely bitter, they took to drinking it dissolved in carbonated water with sugar as 'Indian tonic water'. This was often mixed with gin to help further mask the bitterness and make the medicine more palatable – and so was born the G&T cocktail.[79] (Incidentally, quinine is also the compound within tonic water that glows in the dark when illuminated by ultraviolet light, familiar to those sipping G&Ts in a nightclub.) But it was not until the 1880s, with the mass production of high-quality fever tree bark by the Dutch in Indonesia, that the price of quinine began to decline significantly.[80]

This solved the global quinine supply problem. With the disease barrier that had previously been protecting the continent now crumbling, the gateway opened for colonial expansion into the vast interior of Africa without the abject fear of imminent mortal sickness.[81]

By the early 1880s, the British had already established a presence on the Guinea Coast and in South Africa, and claimed territory around the eastern coastal ports of Mombasa and Berbera (in modern Kenya and Somalia) to secure their sea route to India. France seized a region along the northern bank of the Congo, and Germany claimed Dar es Salaam (modern Tanzania) as well as parts of Togo, Cameroon, Tanganyika and Namibia.[82] In 1884, Otto von Bismarck, the statesman who had united Germany in the previous decade, convened a conference in Berlin to settle disputes between the imperial powers over their territorial claims in Africa – a kind of international diplomatic gentleman's agreement over divvying up the land grab – which is widely considered as the starting gun for the 'scramble for Africa'. Rivalry between the powers escalated the feeding frenzy, with each striving to outdo and gain a strategic advantage over the others. They morally justified the exploitation of the 'Dark Continent', its population and resources as a civilising mission bringing enlightenment and humanitarianism.[83]

Within a single generation, virtually all of Africa was carved up by Britain, France, Germany, Italy, Belgium, Portugal and

Spain, the borders between them bearing no consideration to geography or ethnic groupings.*

The scramble for Africa was facilitated by technological advances in transportation and communication – including steamships, railways and the telegraph – but the crucial development that had enabled Europeans to explore and exploit the African continent were medical countermeasures against the endemic tropical diseases, which until then had made it a death-trap.

So far, we've considered the effects of endemic diseases in different parts of the world. We'll now turn our attention to the raging spread of epidemics through whole populations and see how these sudden, catastrophic jolts of mortality precipitated long-lasting shifts in the affected societies.

* The only countries to remain independent of European rule were the tiny polity of Liberia, created by the relocation of freed slaves from the US and Caribbean, and Ethiopia, which had been able to thwart colonisation by defeating an invading Italian army in 1896.

Chapter 4

Epidemic Disease

When has any such thing been even heard or seen; in what annals has it ever been read that houses were left vacant, cities deserted, the country neglected, the fields too small for the dead and a fearful and universal solitude over the whole earth? Oh happy people of the future, who have not known these miseries and perchance will class our testimony with the fables.
—Petrarch, on receiving word of the death of his beloved, Laura, in 1348, as the plague raged in Europe

The invention of agriculture in several independent locations around the world roughly ten millennia ago has been described as the worst mistake in human history.[1] Farming and living in permanent settlements allow the production of food surpluses and enable higher female fecundity, both of which support population growth, but there can be no doubt that it hasn't been good for human health. The switch from hunting and gathering to cultivating crops and raising livestock certainly led to a contraction of dietary diversity and a higher incidence of nutritional deficiencies; it also required humans to spend more time and energy producing the calories they needed. And there was another unintended consequence of the transition to agriculture – pestilence.

Unlike hunter-gatherers, who may achieve a kill in the wild

and immediately butcher and eat the flesh, pastoralists get a reliable source of meat and hides. And of course, grazing herds kept on pasture are also very effective at transforming plant material that is inedible to humans into protein-rich meat. Animal husbandry also supplies other valuable resources, known as secondary products, that are unavailable to hunters. These include nutritious milk, wool and muscle power: beasts of burden carry heavy packs and pull ploughs, wagons and chariots. But all this involves a much more intimate relationship with animals; people and animals have often lived together in the same abode for warmth, which provided pathogens with a prime opportunity to evolve to jump the species barrier and infect humans. We were given the common cold from horses; chicken pox and shingles from poultry; influenza from pigs or ducks; smallpox and tuberculosis from cattle; and measles from dogs or cattle.[2] Mumps, diphtheria, whooping cough and scarlet fever were also all originally animal diseases that vaulted into human populations.[3] Others seem to have originated in vermin attracted to our food stores and homes, such as leprosy, which comes from mice.[4]*

While some human diseases are believed to be ancient, such as malaria – with which, as we have seen, humanity has had a long evolutionary relationship – most diseases broke into the human population after we adopted agriculture and began to live with domesticated animals.

Agriculture enabled humanity to live in sedentary societies with ever-denser concentrations of people. This close packing of potential hosts provides the perfect conditions for a pathogen to spread rapidly as a crowd disease. What's more, throughout history, towns and cities have often made unsanitary environments,

---

* Diseases can also jump the species barrier into humans from wild animals when our expanding populations encroach on their natural habitats. These are known as spill-over events and account for most of the newly emergent infectious diseases affecting us today, including AIDS, Ebola, Lassa fever, Zika and COVID-19.[5]

with people living among rotting refuse and sewage, thereby contaminating their own water supply. (Hunter-gatherers and nomadic pastoralists, on the other hand, don't face such problems: they simply move on.) All of these developments led to a surge of infectious diseases breaking into the human population.

So while agriculture bequeathed us glittering cities, flourishing commerce and other fruits of civilisation such as writing, these extraordinary gifts came at a price. Like an ancient, prehistoric Faustian pact, the cost of farming and civilisation was the emergence of plagues.

While large outbreaks of disease likely ripped periodically through the cities of the earliest civilisations in Mesopotamia, Egypt, Northern India and China, there is no surviving written record of these first epidemics. And as populations increased across Eurasia, creating ever-denser towns and cities, different areas would have developed their own local set of circulating infectious diseases. But it was when trading networks extended and began to connect the major population centres, ports and entrepôts that pathogens were able to spread far and wide, always finding fresh hosts to infect.

Warfare is invariably associated with the spread of disease. Soldiers drafted from disparate provinces and then packed closely together in large numbers in squalid camps created prime conditions for the mixing and propagation of the pathogens they brought with them. When invading far-flung regions, they were exposed to new, local diseases, while transmitting their own to the native civilians; and when returning from foreign locales, they brought back new diseases to their homelands.[6]

Disease has always exacted a punishing toll upon armies on the march or laying siege, adding to their general exhaustion and malnutrition, but accurate figures only became recorded from the nineteenth century. In the Crimean War of the mid-1850s – to which we'll return in Chapter 8 – Britain lost ten times more troops to dysentery and typhus than to actual combat with the Russians. By the end of the century, in the Boer War fought against Dutch settlers for control of South Africa, the British Army lost five times more men to microbes than enemy action.[7] In fact, the first major conflict in which more combatants were actually killed by each other rather than attendant disease was the Russo-Japanese War of 1904–5 (and was only true for the Japanese). But even in the First World War, when millions of young men were rammed through the meat-grinder of industrialised slaughter on the Western Front, in the eastern theatre both sides still lost more men to disease than to enemy action. It was only by the time of the Second World War, and the arrival of widespread sanitary measures, infection control, vaccinations and antibiotics, that other humans had finally become more of a threat than microorganisms.[8]

The Four Horsemen of the Apocalypse, believed to be punishments sent by God, are Pestilence, War, Famine and Death. The disruption to society caused by war – young men being torn away from their farms to die in some remote land while an invading army plunders stored grain and livestock – often led to food shortages and famine. Malnourished, weakened people, often driven from their homes, are much more susceptible to disease. Throughout history, therefore, war has not only routinely decimated whole armies but spread pestilence and triggered epidemics across the civilian populations. The Thirty Years War, for example, fought mostly within the Holy Roman Empire in the first half of the seventeenth century, saw total military casualties of just over half a million – up to two-thirds of which were caused by disease.[9] But the numbing tragedy of this conflict, and perhaps the greatest medical catastrophe of human history, is that up to 8 million civilians also perished. Here too only a tiny percentage were killed by direct military action; the vast majority succumbed to starvation (12 per cent) or disease (75 per cent) caused by the disruption of the war.[10]

Yet trade has been as important through history in spreading disease and triggering epidemics as war and the march of armies.[11] In the first millennium BC, civilisation in Eurasia passed through a key transition when the numerous, densely populated cities became connected through extensive trade networks.[12] This ushered in the age of devastating crowd diseases and raging epidemics.

The Plague of Athens is the first known epidemic in recorded history, and struck the city in 430 BC, at the beginning of the Peloponnesian War with Sparta. It also spread around the eastern Mediterranean, but with less devastating consequences than in the overcrowded city, where it killed between a quarter and a third of the inhabitants.[13] The plague was recorded by the Greek historian Thucydides, who caught, but survived, the disease and described symptoms including a raging fever, livid patches on the skin, vomiting, severe diarrhoea and convulsions. Athenian society fell apart, as the populace 'became indifferent to every rule of religion or law'[14] and descended into widespread debauchery and crime, believing they were already living with a death sentence.[15]

Plagues would become a recurring feature of history.

PLAGUE OF CYPRIAN

In its heyday in the second century AD, as the capital sat at the heart of a sprawling empire stretching right around the Mediterranean, Rome was particularly prone to the outbreak of epidemics. While it was exemplary at the time for its use of sanitation and for protecting the public water supply from contamination, it was still an extremely crowded city with a population of around one million – the largest in the world at the time. Its traders moved freely across vast areas, and its armies marched to all corners of the known world, creating a network of super-highways for microbes. And all roads led back to Rome.[16] Not only was there the potential for pathogens from far and wide to enter the capital, but the crowded city also provided the perfect conditions for their rapid spread as epidemics.

The so-called Antonine Plague struck the Roman army while battling its age-old rival the Parthians in Mesopotamia at the end of AD 165, and the troops brought it back to Rome the following year.[17] It spread rapidly across Eurasia, reaching as far as India and China, with successive waves recurring until the early 190s.[18] According to the contemporary physician Galen, the symptoms included a scabby skin rash, fever, bloody stools and vomiting,[19] but we don't know the exact nature of the disease: it could have been smallpox, measles or maybe typhus.[20] But we do know it was deadly: the Antonine Plague is believed to have killed between 10 per cent and 30 per cent of the Roman population.[21]

The Plague of Cyprian, named after the Christian bishop of Carthage who witnessed and described it, struck next. Originating in Ethiopia in 249, the epidemic spread across North Africa, through the entire Roman Empire and into Northern Europe, recurring in waves for the next two decades. Again, the causative agent is unclear, but like the Antonine Plague it may have been smallpox or measles, or perhaps a haemorrhagic virus similar to Ebola.[22] The plague claimed two emperors, Hostilian and Claudius II, and about a third of the population of the Roman Empire – possibly as many as 5 million people.[23]

This epidemic is believed to have been a major contributing factor to the so-called 'crisis of the third century', which led to a rapid transformation of the empire between 250 and 275.[24] The financial system disintegrated, and political turmoil destabilised the ruling elite. An army weakened by disease was stretched thin along the empire's lengthy frontiers, unable to repel constant raids and invasions from barbarian tribes or the deep territorial incursions from the rival Sassanian Empire. But perhaps the most significant long-lasting effect of the Plague of Cyprian was the rapid spread of a particular religion.

The mortality and existential crisis caused by the plague made many Romans lose faith in their traditional polytheistic religion, with its pantheon of fractious and devious gods.[25] At the time, Christianity was a relatively obscure and somewhat radical cult,[26] but what made it stand out now was that it preached community-minded charity and care for the sick as a righteous duty.[27] Christian churches throughout the empire responded to the crisis of the plague by encouraging their members to care for those suffering from the disease, even at the risk of catching it themselves. Before the arrival of modern medical practices, nursing – being kept warm and helped to eat and drink – could make a significant difference to an individual's chances of recovery and survival. As a result, the Christian communities would have enjoyed slightly higher survival rates of the plague; but what's more, any pagans who benefitted from Christian charity and were nursed back to health would have felt significant gratitude and commitment to the religion that had saved them. And the prospect of a heavenly afterlife, accessible to those who had lived a virtuous and charitable life, as preached by Christians, surely held a particularly strong appeal during the rampant death of a severe epidemic.[28] While many other institutions were failing, the Christian Church was bolstered by the unfolding devastation of the plague.

This marked the beginning of the rapid spread of Christianity across the empire, despite the fact that its followers were still persecuted for their beliefs by the Roman authorities. Persecution ended when Emperor Constantine issued the Edict of Milan in 313; and in 380, Theodosius declared Christianity Rome's singular state religion. As a result, Christianity would be the dominant religion in Europe and the West for the next 1,500 years

PLAGUE OF JUSTINIAN

By the end of the fifth century, the Western Roman Empire had collapsed under the pressure of invading barbarians, while the eastern half survived as the Byzantine Empire, with its capital at Constantinople. Byzantine culture, scholarship and architecture flourished under the rule of Emperor Justinian I (527–565), and the magnificent Hagia Sophia was constructed. Justinian also rewrote the body of Roman laws, which came to serve as the basis of the Napoleonic Code 1,200 years later, and thereby influenced law-making in much of continental Europe and around the world. But perhaps Justinian's most ambitious project was his bid to reconquer the lost territories in the west in order to resurrect the glory of the unified Roman Empire.

In this, he was largely successful, at least for a time. Justinian's armies successfully crushed the Vandal Kingdom to reclaim North Africa; recaptured the southern portion of the Iberian Peninsula to establish the province of Spania; and conquered the Ostrogothic Kingdom to restore the Dalmatian coast, Sicily and Italy, including Rome itself, to the empire. Justinian also launched repeated campaigns against his powerful eastern neighbour, the Sasanian Empire.[29]

But these successes proved short-lived when biological catastrophe struck. The year 541 saw the arrival of one of the most deadly and feared diseases in human history: the bubonic plague.[30] DNA extracted from skeletons dating to the period confirm that the Justinian Plague was caused by *Yersinia pestis*, the same flea-transmitted bacterium behind the medieval Black Death and mid-nineteenth-century epidemics of plague.[31]

The first plague epidemic is believed to have originated in the Central Asian highlands near the Tibet-Qinghai Plateau and probably travelled with seaborne trade across the Indian Ocean and up the Red Sea.[32] When it reached the Egyptian port of Pelusium it ripped across the busy trade networks around the Mediterranean Sea and throughout the Byzantine Empire, hitting Constantinople in 542.[33] Justinian's enemies weren't spared: the epidemic spread across the Sasanian Empire in Persia, too.[34] The historian Procopius, who witnessed the devastation, wrote, 'During these times there was a pestilence, by which the whole human race came near to being annihilated ... it embraced the whole world, and blighted the lives of all men.'[35]

Between a quarter and half of the population of Constantinople was wiped out within two years,[36] and between 25 million

and 50 million people across Europe and around the Mediterranean perished.[37] With its devastating death toll, the Plague of
Justinian is regarded as the third-worst epidemic in human history, after the Black Death of the 1340s and the influenza
pandemic of 1918.[38]

The initial surge had passed by 550 AD: after raging through
the population, it burned out when it could infect no more susceptible individuals. But fresh outbreaks recurred over many
years – a common feature of epidemics, whereby the disease is
able to sweep back across a region when the acquired resistance
in the population is fading or new generations of susceptible
individuals have been born. The plague reverberated through
history until the mid-eighth century, when the disease finally
receded from Europe, the Mediterranean and the Middle East.[39]

The mass depopulation resulting from the Plague of Justinian
caused significant socioeconomic upheaval across the Byzantine
Empire.[40] Trade around the Mediterranean faltered, leading to
a prolonged period of economic instability.[41] Fiscal records
from the plague years reveal a dramatic decline in tax revenue,[42]
squeezing imperial expenditure, especially on the army.[43] By
588, army pay had been cut by a quarter, and rebellions among
soldiers were breaking out along the eastern borders.[44] The
damage to the empire's finances and military lasted for generations,[45] and before long, the recently reconquered territories in
the west – North Africa, most of Italy, Greece and the Balkans –
were lost again.[46] The decline of the Byzantine Empire following
the recurring waves of plague marked the end of the ancient
world – the classical age of the great Greek and Roman
civilisations – and the beginning of the Middle Ages.[47] The
centre of European civilisation moved from the Mediterranean
rim towards Western and Northern Europe.[48]

The conflict between the Byzantine and Sasanian Empires
continued into the seventh century, further weakening both
sides which had been hit equally hard by successive waves of
plague. This left the region vulnerable to a newly rising power –
the armies of Islam.[49]

Muhammad was born in Mecca around 570, and by the time of his death aged 62, he had united the tribes of the Arabian Peninsula. The Byzantines first clashed with advancing Islamic armies in the late 620s, when Muhammad was still alive, but neither the Byzantine nor Sasanian Empires were able to mount much resistance.[50]

Living as nomads in low population densities, the peoples of Arabia had been less severely affected by the recurring outbreaks of plague than those in the dense towns and cities of the Roman and Persian Empires.[51] After uniting the tribes under a new religion, Muhammad had died in 632, but his successors conquered huge territorial areas of the Byzantine Empire and caused the complete collapse of the Sasanian Empire in 651. By 750, first under the Rashidun and then the Umayyad dynasties, the caliphate had expanded from the Arabian Peninsula to stretch west across North Africa to the Iberian Peninsula, and east throughout Persia to the Indus River. The Islamic empire rapidly filled the power vacuum left by the decline of the two superpowers in the region, which had been weakened by the waves of plague and war with each other.[52]

BLACK DEATH

Eight hundred years after the Plague of Justinian, *Yersinia pestis* returned to western Eurasia with a vengeance – as the Black Death.

This medieval outbreak of the plague may have been triggered by the military campaigns of the Mongols through the Gansu corridor – leading between mountains and desert into the plains of China – possibly as they hunted rodents for meat and fur.[53] It seems that from here, plague-carrying fleas travelled with the Mongol army and then with merchants and their wares along the Silk Roads trade network westwards to the Middle East and the Black Sea, where the epidemic entered Europe. An early instance of biological warfare is recorded from the time.

Caffa (now Feodosia in Ukraine) on the Crimean Peninsula was one of the main trading ports of the maritime republic of Genoa, permitted by the Mongolian khan to operate within his empire. But relations soured, and the Mongol army of the Golden Horde laid siege to Caffa in 1346. Unable to capture the fortified port, the besiegers catapulted over the stronghold walls the corpses of the men who had died of plague breaking out in their camp. The fortress fell, and the fleeing inhabitants are believed to have spread the plague to Europe.[54]

Whatever its exact route, the plague reached Europe in the autumn of 1347, with ships from the east docking at the ports of Sicily, their crews already dead or dying of the strange new disease. The sailors were suffering from high fever, vomiting, acute headache and delirium; and their bodies were marked with protruding, dark boils in their necks, armpits and groins,[55] which not only were exceedingly sensitive but also made strange gurgling sounds.[56] These characteristic boils were due to the plague bacterium infecting the lymph nodes in the body – in fact, the word 'bubonic' derives from the Greek *boubõn*, meaning 'groin'.[57]*

From Sicily the plague passed into mainland Italy and then spread quickly along the Mediterranean coast to France and the Iberian Peninsula, while reaching Byzantium by ship. Florence was hit particularly hard, with the pestilence wiping out 60 per cent of the city's inhabitants. As with other epidemics, the plague was far worse in the cramped, squalid conditions of towns and cities, spread rapidly by rats infested with plague-laden fleas, or perhaps by lice;[59] but rural areas were also severely struck. By spring 1348, the whole of Southern Europe was aflame with the Black Death, and the infection was heading north overland.[60] It crossed the English Channel later that year to strike London, killing almost half of its 60,000 inhabitants.[61] Once infected,

---

* The moniker 'Black Death' didn't appear until centuries after the event. It's possible that it derives from a mistranslation of the Latin *atra mors* – *atra* meaning both 'terrible' and 'black'.[58]

victims often died within a matter of days.⁶² And the plague was
an indiscriminate killer, felling old and young, male and female,
rich and poor, with equal ferocity. So many people succumbed
that the cemeteries overflowed and trenches were dug into
which bodies were piled in mass burials.

By 1353, when it began to recede, the Black Death had spread
its deathly veil across the whole of Europe, North Africa and
the Middle East,⁶³ killing somewhere between one-third and
two-thirds of the population. In all, within just a few years,
some 50–100 million people had died – it would take over two
centuries for populations to recover to their pre-plague levels.
The Black Death was the greatest demographic catastrophe
ever to strike the human race. The influenza epidemic of 1918
may have killed a greater number of people, but by then the
global population had increased hugely. In terms of the sheer
mortality rate among those infected – between 50 and 60 per
cent – the Black Death was by far the deadliest disease in
history.⁶⁴

The short-term effects of the Black Death were profound.
The catastrophic loss of life inflicted unspeakable psychological
trauma on survivors who witnessed the demise of a huge proportion of their community, in many cases including much of
their own families. Society was paralysed, stunned by the apocalyptic event, and survivors gave themselves over to earthly
pleasures, without care for the future. People also suffered from
a profound disruption of economic activity.

But in the long term, as Europe recovered from the colossal
shock to social and economic structures, there were some beneficial outcomes – a silver lining to the storm cloud of bubonic
plague.

Feudalism was the social system in place across Europe in
the fourteenth century. Extensive tracts of land were owned
by a lord of the manor and passed down the family. Rural
peasants were permitted to grow their own crops on parcels
of the land in return for labour, and they also had to serve as
soldiers when the call to arms came from the king. The mass

depopulation caused by the Black Death shook this system to the core.

The deaths of large numbers of the lower classes created an acute shortage of both unskilled labourers on manor lands and skilled artisans and tradesmen in the villages and towns. With labour becoming more valuable, peasants and craftsmen gained bargaining power. Labourers left their manors in search of a better deal. The nobility and governments attempted to control the situation by imposing limits on wages and prohibiting the movement of serfs, but these measures largely failed. Although the restrictions were brought in for the benefit of the feudal lords, it was still in their own best interests to turn a blind eye and accept labourers who were wanting to move – so they were quite happy to poach much-needed workers from their neighbours.[65] With people travelling in search of better work, feudal ties to the manor were weakened.

The extraordinary death rate also meant that large areas of land were now unoccupied and without a surviving owner, and these were transferred to those relatives within the gentry who were still alive. As more land was held among fewer landowners, its value dropped. Plots were sold to peasant farmers who had never owned property before.

Overall, any severe mortality crisis can be expected to lower inequality in society: when labour becomes more scarce, real wages rise, reducing the income disparity between the richest and poorest in society. We have solid evidence that this is exactly what happened after the Black Death.[66] Such a drop in wealth inequality also led to a reduction in capital income inequality, as a larger section of society now had both the means (from higher wages) and the opportunity (due to more land being available, more cheaply) to acquire property of their own.[67] What's more, increased mobility and wages also created a higher general standard of living.

Feudalism was beginning to break down. The system of payment by labour for access to land started to be replaced by receiving wages and paying rent – a shift from indentured serfdom to monetary transactions. This led to a more market-oriented economy[68] and a freer, more mobile society. The Black Death didn't immediately end the feudal system of lord and serf – in England, it didn't disappear until the sixteenth century, and it continued even longer on the continent – but the social conditions created in the aftermath of such a mass depopulation certainly accelerated its demise in parts of Western and Northern Europe.[69]*

As well as empowering individual peasants, depopulation transformed European agriculture. At the beginning of the fourteenth century, Europe's population – although only about one-tenth of what it is today – was pushing right up against the carrying capacity of the cultivable land with medieval crop varieties, tools and farming techniques. Most available land was put to harvesting staple cereal crops, such as wheat, to feed the masses. Market prices for basic foodstuffs were high and the lack of dietary diversity led to malnutrition. Crop rotation systems were often suspended so as to grow much-needed grain even on the fallow fields. This resulted in nutrient-depletion and exhaustion of the soil, and crop yields dropped further. In the years before the arrival of the Black Death, there had been a succession of increasingly severe crop failures – and famines – possibly also linked to a climatic shift to colder, wetter conditions.[72] Much of fourteenth-century Europe was therefore

* Eastern Europe was hit last by the Black Death in 1350–1351, and for reasons which are not entirely clear, suffered death rates only about half as high as the rest of the continent.[70] While it therefore escaped the worst immediate effects of mass mortality, its post-plague trajectory also diverged. The feudal system did not become established in Eastern Europe until after the Black Death, and in fact, the effects of the plague may have contributed to bring about this 'second serfdom' and the long-term worsening of conditions for peasants. The depopulation of Western Europe by the Black Death had stopped the migration of people into the sparsely populated east. Historians have argued that this made aristocratic landlords in Central and Eastern Europe tighten their control over the populace and tie the peasantry to their own estates.[71] Serfdom was not abolished in many of these territories until the early nineteenth century, and it persisted in Russia until the 1860s.

locked in a vicious cycle known as the Malthusian Trap[73] – the population had grown until it was limited by agricultural production and lived in widespread poverty at subsistence level. Before the Black Death, Europe was a stagnant, overcrowded continent.

The Black Death shattered this deadlock.[74] The population collapse meant arable land no longer had to be used to grow just grain to feed everyone, resulting in a diversification of agricultural produce. Food became more plentiful and therefore cheaper for the average peasant and town-dweller, and standards of living rose. Just as significantly, marginal land which had been brought under the plough reverted to woodland or pasture for livestock. Sheep farming is more land-intensive but also more labour-efficient, well-suited to the smaller population – only a handful of shepherds are needed to tend huge flocks of sheep. The growth of the wool industry further invigorated local economies,[75] and wool export transformed the English economy in particular in the late-medieval period.

The long-term historical repercussions of the demographic catastrophe of the Black Death – higher wages, lower food costs, higher living standards, greater social mobility – created a more diversified society and economy in Western Europe. The plague devastated fourteenth-century Europe, but the green shoots of recovery grew back stronger.

This second period of plague in Eurasia remained a threat for the next three centuries before finally fading away again in the late seventeenth. Over this time, the disease flared up in irregular outbreaks, and in fluctuating severity, across different regions. Thus the last great plague outbreaks of 1629–1631 and 1656–1657, for example, hit Southern Europe much harder than the north,[76] killing up to 4 million people in Italy, 2.2 million in France and 1.25 million in Spain. There were less than half a million English victims in these final outbreaks (although the Great Plague of London in 1665–1666 killed almost a quarter of the city's population).[77]

Some historians believe that the Black Death, and subsequent waves of plague, was a contributing factor behind the Western world coming to overtake eastern civilisations such as those of India and, in particular, China from the sixteenth century in terms of economic, technological and industrial development – the so-called 'great divergence'. The argument is that in western Eurasia, the plague established a persistent pattern of high mortality and high income, which created a favourable environment for a series of socioeconomic and political reforms that invigorated faster development and enabled the West to pull ahead. China was less heavily impacted by the plague between the fourteenth and seventeenth centuries and therefore remained locked in the Malthusian Trap, its population living in a state of subsistence at the limits of the carrying capacity of the land.[78]

Still, China was also affected by the plague, with an outbreak of the disease in 1633–1644 believed to have contributed to the fall of the Ming dynasty, which had ruled the empire for almost three centuries. Ming China had already been in decline by the early seventeenth century, but the impact of plague knocked out the last teetering supports. Severe epidemics broke out in Beijing and the region north of the Yangtze. Crops were not planted or harvested, food supplies dwindled and grain prices soared as the commercial economy shut down. With no one able to pay taxes, the imperial treasury ran empty. The government wasn't able to pay the army to quell the peasant revolts breaking out in the provinces or repel the Manchu invaders that attacked the Great Wall. When, in April 1644, a rebel army captured Beijing, the last Ming emperor hanged himself from a tree outside the Forbidden City, and the Manchus then established the Qing dynasty.[79]

GENOCIDAL GERMS

Columbus's 1492 voyage to the Americas marked the start of centuries of European conquest, colonisation and exploitation of the New World. The exploration of the twin continents by the Spanish and Portuguese also triggered a global redistribution of natural resources, known as the Columbian Exchange. Domesticated plants and animals native to the Americas – such as maize, potato, tomato, chilli, tobacco and the turkey – entered the Eurasian diet, while Old World domesticates like wheat, rice, cattle, pig, sheep, chicken and horse were transported to the Americas. But what's more, the Columbian Exchange marked the greatest redistribution of microbes in human history.

After Columbus's first contact, the Spanish continued exploring the islands of the Caribbean and the eastern coastline of the Central and South American mainland, establishing settlements across newly claimed territories. There were rumours of great empires inland – complex civilisations that could supply the gold riches the Spanish yearned for – and privately funded mercenary armies embarked on expeditions into the interior.

Hernán Cortés landed on the Yucatán coast in 1519 and marched on the capital of the Aztec Empire, Tenochtitlan (today's Mexico City), with just sixteen horsemen and around 600 foot soldiers.[80] The Spanish were initially peacefully received, but when they took the Aztec emperor, Moctezuma, hostage they were forced to flee the city, losing many men in the retreat. Hopelessly outnumbered, the Spanish braced themselves for a final, crushing attack – but it never came. Smallpox, brought to the New World by the Europeans, had already begun to rip through the highly susceptible population. The Spaniards returned to Tenochtitlan and laid siege to the city for seventy-five days.[81] When they finally entered the Aztec capital, they encountered a ghost town, with the stockades, houses and streets littered with corpses[82] – victims of the European-brought pestilence.[83]

The disease stalked from village to village through the entire Yucatán region, killing so many that there weren't enough people left alive to farm the fields. Famine followed, and before long, the Aztec civilisation had collapsed. The survivors submitted to Spanish control. By the time the Aztec state had fallen in 1521, smallpox was already spreading into South America along trade routes, soon reaching the Andes and the heart of the Inca Empire.[84]

A decade later, in 1531, another Spanish company of just 62 horsemen and 106 foot soldiers,[85] led by Francisco Pizarro, landed on the coast of Peru to launch an expedition into Incan territory. Smallpox had already wrought its devastation, killing around a third of the population. The emperor too had died, triggering a succession crisis and civil war.[86] Pizarro met no significant military resistance to his invasion and captured the new emperor, Atahualpa, whom he held hostage for eight months, demanding a huge ransom of gold for his safe release. After the gold was gathered from across the empire and delivered to the Spanish, Pizarro had him executed anyway.

The European conquerors had a technological superiority over the warriors of the native civilisations of the Americas. Their bronze weapons, bows and arrows and slings were no match for Spaniards' canons and muskets, cavalry horses and razor-sharp steel swords.[87] But the decisive factor in the clash between the Old and New Worlds was not military but epidemiological.[88] Cortés did not conquer an empire of an estimated six million Aztecs, nor did Pizarro's defeat ten million Inca:[89] both were destroyed by new diseases.

Smallpox had been present in the Old World causing outbreaks for millennia, perhaps from as early as 3,500 years ago in Egypt, India and China.[90] While it utterly decimated the people of the Americas, it still had a deep impact on the history of Europe, despite the population being more resistant. Smallpox struck a young Queen Elizabeth I just four years into her reign, leaving her half-bald and her face pitted with scars, and causing her to rely on wigs and thick cosmetics caked white on her face.[91] Seventeenth-century Europe has been called 'the age of powder and patch' because of the widespread use of white make-up and small dark beauty spots to hide pockmarks.[92]*

---

* While Queen Elizabeth survived her encounter with smallpox, the dynasty that followed her rule, the House of Stuart, was not so lucky. When Charles II returned from France in 1660 to restore the monarchy after the Civil War and Interregnum, he had already lost both his brother and his sister to

Smallpox also killed a multitude of kings, queens and emperors across Europe (as well as the emperors of Japan and Burma),[94] ending dynasties and deflecting the course of succession and disrupting alliances in the French, Spanish, German, Austrian, Russian, Dutch and Swedish royal families.[95]

In the Americas, measles and influenza showed a greater disparity in their impact on the invading Europeans and the indigenous populations. Europeans still died quite regularly of smallpox,[96] but other Old World diseases such as measles and influenza as well as mumps, whooping cough and the common cold were rarely fatal in adults.[97] But for the native Americans who had never previously encountered these diseases, and therefore possessed no genetic resistance or acquired immunity, these diseases could each kill around 30 per cent of those they infected. They raged through the susceptible indigenous populations as so-called 'virgin soil' epidemics, resulting in near-total population collapse.[98] And, as we have seen, when Europeans began transporting African slaves to the Caribbean and Americas, they also introduced the mosquito-borne diseases malaria and yellow fever.[99]

When European explorers ventured deeper into the American interior, they frequently encountered a post-apocalyptic wasteland, with abandoned villages and deserted, overgrown farmland. So extreme was the Great Dying of indigenous American populations that it even caused a blip in the Earth's climate. There is

---

smallpox, and with no legitimate children of his own when he died, he was succeeded by his brother James II. James II, a Catholic, was deposed by the Glorious Revolution of 1688 which saw the crown handed to James' protestant daughter Mary II and her Dutch husband (and cousin) William III of Orange. Mary died shortly afterwards of smallpox, before she had produced an heir. William, who had survived smallpox in his own childhood but lost both parents to the virus, ruled alone until his death, when the crown passed to Mary's sister Anne who became Queen in 1702. When Queen Anne's son and heir himself died of smallpox aged eleven, it triggered a succession crisis. The Stuart line had been wiped out within a generation of Charles II, ravaged by smallpox, and the crown passed to the House of Hanover.[93]

a measurable dip in the atmospheric carbon dioxide levels of the sixteenth and early seventeenth centuries, caused by the abandonment of agricultural land and regrowth of forest over extensive regions – perhaps about 56 million hectares – that created a small cooling effect on the global climate.[100]

The Aztec and Inca Empires were the first to succumb to Eurasian pathogens early in the sixteenth century, when the Portuguese conquest and colonisation of Brazil was also aided by Old World diseases. A hundred years later, they began devastating indigenous tribes further north as well. When the Mayflower arrived at Cape Cod in 1620, savage epidemics of pathogens imported by earlier European explorers had raced ahead of the Pilgrims,[101] so they were under the impression that they had set foot on a largely vacant continent, full of fertile but uninhabited land ready for the plough. This would feed into the nineteenth-century concept of the Manifest Destiny – that the westward expansion of the United States was not only justified but inevitable, as settlers populated a largely empty continent that seemed to have been just waiting for them.

By the end of the seventeenth century,[102] the full panoply of Eurasian diseases was also endemic in the Americas and constantly swirling around the population, present as the background burden of diseases people are exposed to in their childhood, as they had in the Old World.*

---

* Columbus and his crews were not actually the first voyagers from the Old World to visit the Americas. Half a millennium earlier, in the late tenth century, Norse sailors had ventured across the far North Atlantic from Scandinavia to settle Iceland, then Greenland, and finally a place far to the west they called Vinland.[103] The archaeological remains of a Norse settlement have been identified on the North American mainland, at L'Anse aux Meadows on the northern tip of Newfoundland, and recently carbon-dated to around AD 1021.[104] These Norsemen made contact, and encountered hostile relations, with indigenous tribes, which they referred to as Skrælings. But they don't appear to have introduced any epidemic-causing diseases, despite the fact that smallpox, for example, was already widespread in Northern Europe by the Viking Age.[105] The combination of a small source population in Iceland, combined with the cold conditions in the open-top longboats

It may be impossible to ever determine an exact figure for how many people in the Americas died of epidemics as they were first exposed to Eurasian pathogens. The estimated overall population loss rates are contentious but range from 40 per cent[106] to as much as 95 per cent,[107] with most recent calculations coming in towards that upper end.[108] What seems likely is that the pre-contact population of the Americas in 1492 was around 55–60 million people, which crashed to just over 5 million by 1600.[109] Even with the influx of settlers, first from Europe and then from the rest of the world, and the arrival of Africans through the slave trade, it took around three and a half centuries for the population of the Americas to recover from the microbial massacre.[110]

I have focussed here on the devastating effects of Old World diseases upon the susceptible populations of the Americas, but the contact with European explorers and colonists was just as bad for other previously isolated human populations, such as the Aborigine people of Australia, the Māori peoples of New Zealand, the Khoisan of South Africa and the native inhabitants of Pacific islands such as Fiji.[111] Remarking upon the catastrophic decline of indigenous populations after contact, Charles Darwin wrote in his journal in January 1836, 'Wherever the European has trod, death seems to pursue the aboriginal. We may look to the wide extent of the Americas, Polynesia, the Cape of Good Hope, and Australia, and we find the same result.'[112]

An important question arises from this. When contact was first established between the Old and New Worlds, why were New World populations devastated by Old World diseases, but not vice versa? It seems that, in this respect, the Columbian Exchange was one-sided.*

---

during the crossing, had presumably served to sanitise the crew of such infectious diseases.[106] The Norse American adventure had no meaningful repercussion for either the Norse, the rest of Eurasia or the Americas.

* There is one major disease, however – though nowhere near as lethal as the smallpox, measles and influenza epidemics were in the Americas – that may

Humans migrating out of Africa and dispersing around the world reached North America around 15,000 years ago by crossing the Bering land bridge.[116] This was a wide corridor of ocean floor, exposed as dry land by the very low sea levels during the last ice age, which linked Siberia and Alaska. From here, they spread south to the Isthmus of Panama and crossed into South America. By around 11,000 years ago, as sea levels rose again with the thawing of the world, the land bridge disappeared beneath the ocean waves and the eastern and western hemispheres of the Earth became biologically isolated from each other.[117] The small group of humans that had crossed into Alaska did so before the development of agriculture and the domestication of animals (with the exception of the dog), and therefore before the existence of the many crowd diseases that subsequently emerged in Eurasia. (Other, more ancient diseases such as malaria were not able to cross the freezing Bering land bridge with these first migrants.) So when Eurasia and the Americas became separated again, the human world in the Americas was essentially free of infectious diseases. And significantly, the American populations didn't go on to develop crowd diseases of their own. They too domesticated wild plant species

---

have been indigenous to the Americas and transferred to Eurasia after first contact. The first reported outbreak of syphilis in Europe was immediately after Columbus's return from his initial voyage, during the French siege of Naples in 1493. The presumption is that mercenaries fighting in the French army had served with Columbus's transatlantic expedition and got infected through sexual contact with indigenous people.[113] Within just a few years, this sexually transmitted infection had spread around Europe and then the world, the name used in each country reflecting national rivalries: Italians called it 'the French disease', the French 'the disease of Naples', Russians 'the Polish disease' and Poles 'the German disease'. Across the Middle East it was referred to as 'the European pustules', in India 'the Franks' and in Japan 'Tang sore', referring to the Tang dynasty of China.[114] It should be noted, though, that this 'Columbian theory' for the introduction of syphilis into Eurasia has been facing mounting criticism in recent years after the uncovering of evidence suggesting the presence of syphilis in European populations before 1492.[115]

as crops and developed agriculture and civilisation, but there were very few large animals they could domesticate.

The Aztecs and Incas created sophisticated civilisations that spread across extensive areas, with well-developed transport networks, administration systems and dense urban centres. Indeed, in the early sixteenth century, the Aztec capital Tenochtitlan was among the most populous cities in the world – five times larger than London and on a par with Paris, Venice and Constantinople.[118] These New World civilisations would have been as conducive to raging epidemics as the Roman Empire or medieval Europe. There just weren't any crowd diseases in the pre-Columbian Americas.

This is not to say that before the arrival of the Europeans indigenous Americans lived in some kind of pathogenic Garden of Eden – they still suffered dysentery, intestinal parasites, insect-borne afflictions including Lyme disease[119] and tuberculosis.[120] But they were spared the scourge of widespread plagues as they'd never lived in close association with large numbers of domesticated animals.[121]

The Americas, with large populations and no natural immunity to Old World diseases, were like a wide expanse of dry woodland when just a few sparks of disease brought on European ships ignited blazing forest fires that raged across the land.

THE TRANSATLANTIC SLAVE TRADE

While the introduction of Old World diseases had catastrophic consequences for indigenous Americans, the response of the European colonists to this great depopulation claimed more victims. The same epidemics that had made the Americas so unable to repel European conquest had also left the colonists bereft of a local population that could be forced into labour on their plantations and mines; so they turned to Africa.

The first slaves transported to American colonies were taken by the Spanish to Hispaniola in 1502 to work the newly established tobacco and sugar plantations and dig mine shafts to find hoped-for gold seams. They arrived from Spain where they had previously been working and had converted to Christianity. When it became apparent that more labour was needed for the economic exploitation of the colonies, the Spanish king, Charles V, ordered in 1518 the direct transport of slaves captured from the West African coastline, marking the beginning of the transatlantic trade route.[122] Before long, the concept of race was invented and then reinforced as a colonial construct to dehumanise black Africans and justify seizing 'others' as chattel and enslaving them into forced labour.

Demand for African slaves increased with the expansion of the colonies and plantations, and especially after insect-borne diseases such as malaria and yellow fever had become established in the Caribbean islands and tropical regions of North and South America around the mid-1600s.[123] European colonists, as well as indentured labourers from their own homelands, were as susceptible to these tropical diseases as indigenous Americans. As we saw in the last chapter, African adults were seasoned against yellow fever and malarial infections and had evolved genetic adaptations to malaria in particular, such as sickle cell and Duffy negativity, that provided resistance. Without understanding the intricacies of the immune system or these red blood cell mutations, Europeans were nonetheless well aware of how Africans appeared hardened against the tropical diseases to which they themselves so helplessly succumbed.[124] So unlike in the temperate latitudes of North America (and later Oceania), where migrants from Europe could be attracted to settle and farm, the colonies in the tropical Americas became dependent on imported African labour for their plantation economies. The market price for different labourers reflected their resistance against tropical diseases. Slaves imported directly from Africa were worth three times more than an indentured European labourer and double an indigenous slave; and African slaves who had already been proven to be resistant to the local diseases were worth twice as much as a fresh import.[125]

The produce cultivated on these plantations was primarily cash crops that could be traded within or between colonies or fetch high prices back in Europe – sugar, tobacco, tea, coffee and, later, cotton.* Unlike staple crops such as wheat, rice or potatoes, these plants were valued not for their ability to feed populations but for other effects on our bodies and brains. We will return to our biological cravings for these often addictive substances in Chapter 6.

1918 INFLUENZA PANDEMIC

In terms of the total number of people killed by a single epidemic, the global influenza pandemic that emerged in 1918 was the greatest killer since the Black Death, and possibly in the whole of human history.[126] But it is often discussed only briefly in history books as an endnote to the First World War. With the world already numbed by the horrors of the Great War at the time, the flu pandemic has been described as 'a global calamity the world forgot'.[127]

It's not known where the outbreak originated, but the first documented cases of an unusual respiratory fever were in an army camp in Kansas in March 1918.[128] It was being reported in France by April, possibly spread there by soldiers of the American Expeditionary Forces arriving in Brest, the main port receiving troops from across the Atlantic.[129] Within a month, *grippe* had broken out in the French lines, as well as among British soldiers at the camp in Étaples. The symptoms of this new respiratory illness included a high fever, racing pulse and the coughing up of blood. This developed into breathlessness

---

* Cotton production in the southern states of the US did not become a major activity until the industrial revolution in Britain and invention of the cotton gin. This early industrial device greatly increased the speed with which the cotton fibres could be separated from the seeds, and so drove greater demand for the raw material.

and a distinctive dusky-blue discolouration of the face as the body was starved of oxygen. Autopsies revealed swollen lungs full of thick yellowish pus and haemorrhaged blood – the victims had drowned in their own internal fluids.[130] Because of wartime secrecy, the world first heard about the new disease when it spread through neutral Spain in May 1918, where it triggered public alarm and made headline news. The pandemic therefore came to be known, rather unfairly, as 'Spanish flu'.

The influenza pandemic started off relatively mild, but successive waves became progressively more serious and spread further, with the virus evolving to be more virulent in its human hosts. During the months immediately before and after the end of the war, this influenza strain was killing ten times more infected people than normal flus.[131] But the 1918 influenza pandemic was exceptional for another reason. Most flus tend to kill the very young and the old – those with weak immune systems – and so they present a U-shaped death curve. What was odd about the 1918 influenza was that the disease had a pronounced W-shaped death curve: it also killed huge numbers of people aged between 20 and 40 – young adults in their prime.[132] Why this might have been is something of a mystery. One possibility is that in the healthiest individuals, the body's immune system grossly overreacts to the infection in an intense autoimmune response – known as a cytokine storm – that causes extensive lung damage leading to death. Another hypothesis, nicknamed the 'original antigenic sin', relates to the way the immune system remembers a pathogen it has previously been exposed to, which then limits how effectively it responds to infections of slightly different variants of the virus later in life.[133] The war-time conditions possibly also contributed to the abnormally high mortality of middle-aged people, including crowding in army camps and factories.[134] Overall, around half of the victims of the influenza were young adults aged between 20 and 40.[135]

The war didn't cause the 1918 influenza – the spillover of the virus from birds into the human population would have happened anyway – but the troop movements and disruption of the war certainly contributed to its rapid spread around the globe. And it's possible that the trench environment enhanced the extraordinary virulence of this strain. Normally, a pathogen adapts over time to moderate its virulence and keep its host alive longer, so as to be transmitted to more people. But in the conditions of the trenches, where soldiers were hunkered down for weeks at a time in unnatural proximity to one another, with their death rate from other causes already very high, there would have been less selective pressure on the virus to reduce its virulence; and it thus achieved much higher death rates among those infected. Soldiers immobilised by the disease continued to transmit it to those around them, and when those afflicted by the worst variants were brought out of the trenches to a crowded field hospital, it spread among the other wounded and medical staff. Fresh soldiers entering the trenches all the time ensured that the virus always had access to new susceptible hosts.[136]

The end of the war and demobilisation of troops occurred in the middle of the second wave of the pandemic, in November 1918. Soldiers returning to their families around the world were greeted by packed homecoming parties in the streets, which enabled the virus to spread even more widely.[137] A third and less lethal wave rippled around the world in early 1919, before the pandemic finally receded.

All in all, the 1918 influenza pandemic is thought to have infected around 500 million people – one-third of the entire global population[138] – killing at least 50 million[139] and perhaps as many as 100 million[140] of them.[141] Most of the deaths occurred in just the few weeks between mid-September and mid-December 1918. In terms of mortality, this influenza pandemic possibly surpasses the First and Second World Wars combined.[142]

The 1918 influenza caused a huge loss of life, but, like the other epidemics we've looked at in this chapter, could this fast-spreading, virulent disease have affected historical events?

THE END OF WWI

When Russia withdrew from the war a year after the Bolshevik Revolution, Germany was able to move many of its forces away from the east. Fifty divisions of experienced troops and 3,000 guns were redeployed to the Western Front,[143] so, by April 1918, the Germans had a vast advantage of 324,000 riflemen.[144] In several sectors along the Western Front, German forces outnumbered the British and French four to one.[145] In March, the German Supreme Army Command launched a new offensive, known as *Kaiserschlacht* or 'Kaiser's Battle', in an attempt to secure a decisive victory against the Allies before fresh American troops could be fully deployed and turn the tide. The plan was to break through the lines with fast-moving and specially trained stormtroopers to outflank and defeat the British Army, and so force the French to sue for peace.

At first, the spring offensive seemed to succeed. They were able to advance more than 60 kilometres into northern France. When their artillery, now in striking distance of the French capital, began shelling Paris, more than a million Parisians fled.[146] But by June, the phases of this spring offensive were faltering.

Logistical problems played a role, but it seems likely that the arrival of the influenza pandemic hit the exhausted German troops harder than their Allied counterparts. German soldiers were malnourished as a result of the Royal Navy's successful blockading of food imports, and so their resistance to the disease may have been lower than the comparatively well-fed Allied soldiers.[147] The flu had also struck the German army three weeks earlier than the Allies, appearing in March and peaking in early July. Although this first wave of the pandemic was much less lethal than the second wave that came in the autumn, every soldier who succumbed was still laid up for several days, and many were left debilitated afterwards.[148]

Although writing after the war, and largely attempting to exonerate his own failings, the architect of the spring offensive,

General Erich Ludendorff, blamed the flu for sickening his men and lowering their morale during the critical period of the Kaiserschlacht: 'It was a grievous business having to listen every morning to the Chiefs of Staff's recital of the number of influenza cases, and their complaints about the weakness of their troops.'[149] Divisions of 10,000 soldiers were reporting up to 2,000 cases of influenza, and German commanders would only know their unit strength on the day of an attack.[150] Over the summer, between 139,000 and half a million German troops were temporarily incapacitated with the sickness.[151]

Conditions on the home front in Germany were also dire. Severe food shortages due to the British blockade caused an estimated 424,000 excess civilian deaths during the war, compounded by scarcities of coal and warm clothing. A further 209,000 deaths were caused by the flu pandemic – the victims no doubt already weakened by hunger.[152] Morale was collapsing within both the military and the civilian populations, and more and more Germans began demanding an end to the war. Many soldiers, utterly demoralised from fighting on the front line, used the flu as a pretext to go AWOL.[153]

As the strength drained from the German spring offensive, the Allies were only gaining in number as fresh American troops flooded across the Atlantic and onto the Western Front. By July, the numerical advantage had swung back in the favour of the Allies. The following month, their counteroffensives pushed the Central Powers back, undoing the gains made during the spring offensive, before breaking through the German lines in October. By the time the deadlier second wave of the pandemic hit in the autumn, the German army was already beaten. With crushing defeats on the battlefield, and revolution at home, the kaiser was forced to abdicate, and the new Weimar Republic called for an armistice on 11 November.

While influenza further demoralised and weakened German forces, contributing to the capitulation of the German Empire, the horrors of the pandemic brought about unity in another great country, with India able to demonstrate its strength in self-rule.

INDIA

India was hit particularly hard by the 1918 influenza pandemic, with between 12 million[154] and 18 million[155] people estimated to have been killed – more than the total number of combatants killed during the war.

The pandemic arrived in India in late May 1918, rippling out of the western port city of Mumbai. As with the pattern found in the rest of the world, this first wave was not particularly worrisome, but when the second, far more virulent variant broke out in September, the results were utterly devastating.[156] Death swept through cities and rural villages alike – burial grounds and burning ghats were overwhelmed with bodies. The north and west of India were hit earliest and hardest, with death rates of between 4.5 and 6 per cent. India's colonial masters, however, were largely insulated from the worst of the devastation. The British were able to take refuge in their large houses, with staff to tend to them if they fell sick, or retreat to their hill stations in the cooler highlands.[157] As a result, they experienced a death rate eight times lower than that of Indians living huddled together in unsanitary conditions. The viral pandemic respected privilege.

The inability, or perhaps even negligence, of the British colonial rulers to mitigate the colossal death toll was perceived as one more injustice by those advocating independence. The effects of the second influenza wave in India had been exacerbated by drought and food shortages, but despite an unfolding famine, food grown in India was still being shipped to support the British war effort in Europe. It was clear that the British rulers were prioritising their own interests above the needs of the Indian people.[158]

Activists of the Indian independence movement – already working in local communities across the caste divides – organised relief centres, distributed herbal medicines and other supplies, and arranged the removal of the dead. Such grassroots efforts and organisation had existed for a while, but the influenza pandemic united them across the country for a single cause. Local leaders were doing their best to address the health crisis that the Raj government seemed to be largely ignoring.[159]

When the war in Europe ended in November 1918, nationalists hoped for concessions towards much greater Indian autonomy, especially after the huge sacrifices that one million Indian soldiers had made for the British Empire. The Secretary of State for India, Edwin Montagu, had implied in 1917 that India could soon progress to self-government like Canada and Australia, but the actual reforms offered in 1919 shattered this expectation.[160] Furthermore, the Rowlatt Act of March 1919 extended the emergency powers granted during the First World War to indefinite imprisonment without trial for sedition,[161] effectively continuing martial law in peacetime India. Indians had been expecting more freedom; instead, they were slapped with further repression.

As tensions rippled across the country, one figure rose to prominence in directing the mounting anti-colonial sentiment into effective civil disobedience. Mohandas Gandhi had returned to his homeland in 1915 from South Africa, where he had been working as a civil rights activist, on the urging of a leading politician of the nationalist party, the Indian National Congress. After recovering from a prolonged illness – which may have been influenza during the peak of the second wave – Gandhi called for non-violent resistance against British repression.[162] The bubbling discontent and social agitation erupted in spring 1919 with a series of strikes and protests. In the city of Amritsar in the Punjab on 13 April 1919, British soldiers opened fire on a crowd of peaceful protestors, killing hundreds.[163]

The neglectful mismanagement of the pandemic response, the Rowlatt Act, the Amritsar massacre – all these outrages played into the anti-colonial rhetoric and helped to unite India behind the demand for independence. By 1920, Gandhi had started the non-cooperation movement, urging all Indians to boycott British courts and educational institutions, resign from government positions and refuse to pay taxes. The independence movement had been going for years, but activists like Gandhi could now rely on widespread grassroots support. The virus, and the Raj's failings in dealing with it, had mobilised communities across India to help the sick; they were now organised to fight for independence. Even so, it took almost three more decades and another world war for India finally to be free of colonial rule.[164]

Chapter 5

Demographics

Population size is the most important determinant of national power. With it, a lack of other determinants of power can be overcome. Without it, great power status is impossible.

—A.F.K. Organski, *World Politics*

We've already looked at the fundamentals of human reproduction, and how the evolution of big brains and bipedalism drove the development of pair-bonding, romantic love and the family. In this chapter, we'll look at the bigger picture: how human populations shaped our history.

Compared to our closest evolutionary cousins, the other species of great apes – chimpanzees and bonobos, gorillas and orang-utans – humans have a remarkably slow life history. We saw in Chapter 2 that human children need a long period of development after birth before they no longer depend on their parents for survival. Even after infancy, humans take a long time to reach sexual maturity and begin their own reproductive lifetime. This period of adolescence – our teenage years – is far longer in humans than any other great ape. In foraging societies, on average, females start to reproduce at the age of 19, compared to 10 in gorillas and 13 in chimpanzees.[1]

While we are very slow to start reproducing, humans are exceptional in how frequently females can give birth. The average birth interval in traditional hunter-gatherer societies is around three years – significantly shorter than the four-, five-and-a-half- and eight-year intervals for gorillas, chimpanzees and orang-utans, respectively.[2] Our ape relatives are therefore caught in a demographic dilemma: their birth rates are barely greater than their death rates and so their populations can grow only very slowly.[3] Not only do humans have a comparatively huge reproductive potential, but human babies are much more likely to survive to reproductive age than their chimpanzee cousins.[4] So what are the secrets of our demographic success?

As we saw in Chapter 2, humans are a species of persistent cooperators. This is also apparent in the way we breed. Humans raise their offspring within a social system known as cooperative breeding, where members of the group care for children that aren't their own. As a reproductive strategy, this is pretty rare in nature: in only about 3 per cent of bird or mammal species do individuals other than the parents help raise the young. It occurs in some primate species, but in none of the great apes other than us humans. In traditional human societies, childcare is shared among siblings, as well as the wider family. In particular, as human females have great post-menopausal longevity, grandmothers past their own reproductive lifetime are able to support the younger generations of their family.

Humans are generous not just with their time lending a helping hand with childcare, but also in sharing food. Mothers with a young baby are hit with a double whammy. While still nursing they not only need a higher calorie intake to support the breast-feeding but also don't have enough time or energy to forage for food (to provide this additional nutrition for themselves and their older children). Food sharing therefore makes a huge difference to her reproductive capability. Adults also readily share food with the juveniles in the group who aren't yet able to forage enough to be self-sufficient. In other apes (and indeed most other mammal species), parental care stops with weaning; after that a juvenile becomes independent and is responsible for finding its own food. Humans developed a different strategy: infants are weaned early, but then their food intake is effectively subsidised for several more years by those around them. In traditional hunter-gatherer societies, a human baby is weaned at the age of three – but already at the age of one in industrialised societies, aided by dietary supplements and bottle feeding.

And it's not only other adults, such as the father or grandmother, who contribute extensively to supporting the mother with her child-rearing. Older children help support their family by contributing foods that are more easily available, such as fruit or small animals, or helping out with the less strenuous farming tasks in agricultural societies. And they assist with the crucial processing of raw food before it can be consumed – which helps make the nutrients more digestible or preserves them for longer – such as pounding, soaking and cooking. Or they collect firewood and fetch water, which are chores that can take several hours every day. Thus by the time a mother has two or three older children, the task of raising further offspring actually gets easier.

Compared to our ape cousins, who can raise only one child at a time to independence, cooperative breeding, food sharing and not least biparental investment in childcare means human mothers are able to increase their reproductive output by stacking up a chain of overlapping juveniles, relying also on the support of the elder children to help feed and support the youngest. And it is for this demographic superpower – the rate at which we can grow our population – that humans are an absolutely remarkable ape.[5]

But while we exhibit a much greater reproductive growth rate than other great ape species, there have been profound differences in demographic potential *between* different populations of humans. Most notably, the development of agriculture over ten thousand years ago created an enormous shift in human demographics.

For most of the history of civilisation – before the advent of modern machinery, artificial fertilisers, pesticides and herbicides – agriculture invariably demanded back-breaking labour. Agriculturalists had to clear forest to create farmland, manage the land by ploughing or maintaining irrigation canals, and then sow their fields, nurture the crops and weed out other plants. The harvest of cereal crops requires reaping, threshing and winnowing, before storing the grain and finally milling it for consumption. Pastoralists keeping livestock were able to exploit natural plant growth for grazing but still had to invest a lot of time herding their animals, defending them from predators, and preparing fodder for the winter.[6]

Agriculture also exposes its practitioners to a number of health problems and subsistence risks. Eating a lot of grains creates more wear and tear on our teeth, and the focus on carbohydrate-rich foods promotes dental decay.[7] There is also a higher risk of malnutrition from relying on a small range of domesticated plants and animals. Even more acutely, while hunter-gathers are flexible eaters, roaming over wide areas to access a great diversity of edible wild plants – including roots, tubers, berries and leaves – at different times of the year, a farming population quickly faces famine if its staple crop suffers harvest failure. And as we saw in Chapter 4, agriculturalists are more exposed to transmissible diseases from living in close association with their livestock, and cheek to jowl with large groups of other people.[8]

But all that being said, farming can yield a greater return of food than foraging. All the effort invested in tending the land and crops means that an acre of farmland provides much more nutrition than the same acre of forest or wild grassland, and herding livestock is a more efficient way of deriving meat and other animal products than hunting wild game. This agricultural surplus can be stored from one year to the next and also free up labour for other occupations not focussed on food production, such as specialist craftspeople, leading to the development of more complex technology and social organisation.[9]

It has been generally accepted that agricultural societies can support a faster birth rate than hunter-gatherer societies, and so their populations grow more rapidly. It has been argued, for example, that sedentary farming societies can have babies in quicker succession because they don't have to carry their infants long distances, as do forager groups roaming between different camps. More recent archaeological studies, however, have suggested that the long-term population growth rates of some hunter-gatherer societies could match that of prehistoric agriculturalists in both the Old and New Worlds.[10] But this is only up to the carrying capacity of the land, which is lower with a hunter-gatherer lifestyle than an agrarian one.

So even if agricultural food production doesn't necessarily sustain a faster population growth rate, the fact that farming yields more nutrition from the same area of land means that it can support greater total population numbers. And when a particular area becomes too crowded with arable farmers or herders, and the carrying capacity of the land is maxed out, large numbers of people begin dispersing outwards to settle surrounding areas. The irony of agriculture, with its sedentary lifestyle rooted to a particular patch of land, is that the demographic pressure created by increasingly dense populations pushes farmers ever onwards to displace hunter-gatherer communities.

THE BANTU EXPANSION

Africa is truly enormous. It covers an area almost as big as the continents of Europe and North America combined, with a huge variety of environments and peoples – indeed, there is more genetic diversity among the people of Africa than within the whole rest of the human species around the globe. This is a consequence of our long evolutionary history within the continent, before a small population migrated out of Africa during the last ice age and populated the rest of the world. It's all the more surprising, then, that sub-Saharan Africa is astonishingly uniform when it comes to the languages spoken; this vast continental region is much less linguistically diverse than Asia or pre-Columbian America.

The majority of sub-Saharan Africans today – more than 200 million people – speak a tongue that is part of a tight-knit group of around 500 closely related languages known as the Bantu family.[11] This particular language family appears to have emerged around 5,000 years ago in a region of western Central Africa that now straddles the border between Nigeria and Cameroon, from where it spread rapidly to cover most of the central, eastern and southern areas of the continent. Separate languages gradually developed and transformed as they propagated – shifting in their pronunciation, vocabulary and grammar – so that today we find a branching family tree of Bantu languages draped across the sub-Saharan continent, with the stem rooted in tropical West Africa.

The rapid spread of this language family across sub-Saharan Africa is known as the Bantu Expansion. Its staggering scale is perhaps best grasped by comparison. The Bantu family is just one of 177 language subgroups within the wider phylum of Niger-Congo languages.[12] It is equivalent to, for example, the Nordic (or North Germanic) branch of languages – spoken in Denmark, Norway, Sweden and Iceland – within the whole Indo-European language phylum that extends today in two broad swathes: across Europe and Russia; and from Iran through to northern India.[13] But whereas the Nordic languages are only spoken in one small corner of the sprawling patchwork quilt of the different tongues of Europe, the Bantu family extends across 9 million square kilometres of sub-Saharan Africa.[14] It's as if the whole of Europe spoke close variants of the same language.

So, you may wonder, why did so many people across such a huge area all come to speak the same small family of languages? What drove the rapid expansion of this language family? Did the original Bantu-speakers perhaps exhibit military prowess that enabled them to drive a determined campaign of conquest across the continent[15] (akin, for example, to the cause of Spanish prevalence in Latin America today)? Or if the interaction between populations wasn't violent, did a large group of incoming Bantu migrants completely replace or interbreed with a pre-existing population? Or did the Bantu languages spread not through population movement but by a process of cultural diffusion, whereby a custom or technology is progressively transferred from one group to its neighbours?

Sometime around 4,000 years ago, Bantu-speaking peoples began dispersing out of their homeland on the border between modern-day Nigeria and Cameroon.[16] This migration proceeded along two main routes. One branch headed south from the ancestral lands before bearing eastwards across the equatorial rainforest into East Africa, and then down towards the southern tip of the continent. By about 2,500 years ago, Bantu-speaking people had reached Lake Victoria in East Africa and mastered iron toolmaking technology,[17] before expanding further inland.[18] A second major migration pathway bore southwest along the coastal plains,[19] and by around 1,500 years ago, Bantu-speakers had spread as far as Southern Africa.[20] Genetic studies indicate that this expansion didn't proceed as one smooth, continuous diaspora, but in fits and starts in successive phases of expansion,[21] like ripples of migration lapping over one another.

What's more, when the Bantu arrived in new locations, they not only brought with them their way of speaking but also introduced new technologies. The earliest stages of the expansion through the equatorial rainforest are associated with the proliferation of villages and sedentary living, as well as the appearance of pottery and large polished stone tools – in particular axes and hoes.[22] After this initial phase, the Bantu Expansion also spread agriculture and metallurgy. The Bantu-speaking cultures cultivated a set of very productive crops, including staple cereals such as pearl millet, as well as pulses, yams and bananas; and they kept domesticated guineafowl and goats,[23] all of which enabled them to support dense populations.*

---

* The earliest phase of the expansion occurred before the first archaeological evidence of agriculture, but it's likely that these Bantu-speaking migrants did practise the paracultivation – the management of wild plants in their natural environment – of wild yams.[24] Evidence has also emerged that a

The genetic analyses of populations across sub-Saharan Africa have illuminated the past movements of the Bantu peoples during these major migrations.[26] Most notably, within Bantu-speaking populations today there's a great deal less diversity in the Y-chromosome DNA, which is transferred solely from father to son, than there is within the mitochondrial DNA passed from mother to children. This suggests that the Bantu languages were spread mostly by migrating men, who then interbred with indigenous hunter-gatherer women – possibly in polygynous relationships.[27] And there wasn't just mixing of genes between the indigenous populations and the incoming Bantu migrants; some aspects of local hunter-gatherer languages were absorbed into Bantu linguistics. For example, several Bantu languages in Southern Africa have adopted the consonants made of clicking sounds from the Khoisan tribes.[28]

What has become clear in recent years, therefore, is that the Bantu Expansion was not just the diffusion of languages, but the spread of an entire cultural package – language, agricultural lifestyle and technologies – carried by the physical migration of Bantu farmers. Farming produced an abundance of food, pottery allowed surpluses to be stored and cooked, and iron tools made possible the more effective clearing of the land. A stationary lifestyle supported by crops and domesticated animals allowed for greater population that drove the expansion onwards into new territories occupied by hunter-gatherers.

Thus, the astonishing extent of the Bantu languages across Africa today is the conspicuous relic of one of the most dramatic demographic events in human history.[29] Other large-scale language dispersals have occurred in the last 10,000 years, such as the spread of Austronesian-speaking farmers through

climate-induced reduction of the rainforest in western Central Africa around 2,500 years ago may have encouraged the initial Bantu migration.[25] The majority of the spread, however, is believed to have been powered by the demographic drive of growing Bantu populations feeding themselves with agriculture.

Polynesia and Micronesia, but the Bantu expansion is utterly remarkable for its sheer scale and pace. The second stage of the Bantu Expansion, over 4,000 kilometres from central Cameroon to Southern Africa, took less than two millennia – a staggering rate for population dispersal.[30, 31]

After the domestication of wild plant and animal species around the world over ten thousand years ago, agricultural societies expanded into new territories, interbred with the native populations and replaced the original hunter-gatherer lifestyle with arable farming or pastoralism. Ultimately, farming came to displace foraging over much of the world, largely driven by demographic factors – farming supported greater population densities (and perhaps a more rapid growth rate) and powered expanding migration waves. As agriculturalists spread, they carried with them not only their agricultural way of life, but also languages and technologies.

MILITARY MIGHT

Once non-nomadic agriculturalism and then cities became established, demographic trends continued to influence the history of civilisation and the battle for power between settled states. For a while, small states such as Venice and the Netherlands were able to break the dependence of both economic and military strength on population numbers by operating lucrative maritime trading empires and paying mercenaries to fight their wars. But in general, the military strength of states has been directly tied to their population size and the number of fighting-age men.

This is not to deny that other factors play a role in war – the training of troops, the terrain of the battlefield or the tactical brilliance of a general. Sometimes, a new military technology serves as a 'force multiplier', increasing the effectiveness of one side's soldiers on the battlefield and tipping the balance. At times, such innovations have shattered the equilibrium between powers, resulting in the collapse of long-established civilisations and empires and a reconfiguration of the world order. But invariably, any new technological advance – whether that is the chariot, the iron blade or the gunpowder firearm – and the tactical advantage it bestows, is before long acquired by competitors, and the balance of power is restored. Armed conflicts once again become primarily determined by the weight of numbers.

Across the millennia, the key to military success was simply to command greater forces on the battlefield. There have been exceptions to this rule of course, when a smaller force was able to achieve victory against the odds. Such David and Goliath encounters include the Battles of Marathon (490 BC) and Agincourt (AD 1415), and more recently the Six-Day War (1967). But underdog triumphs are few and far between in history.*

It goes without saying that larger populations yield larger armies. And so by extension, the more populous states tend to prevail with the outbreak of armed conflict, either as the invaders or defending their own boundaries against aggressors. Smaller states tend to get overpowered and absorbed by larger states. And of course this process can snowball. The state able to field the larger army defeats its neighbours and brings those territories under its own dominion. With control over more land and more men, the stronger state can conquer and absorb more and more adjoining territory. Provided this state is able to maintain internal stability and supply a swelling army under one banner, it can continue to expand across larger and larger regions. Thus is the birth of empires.

From the days of clashing tribes, through the civilisations of antiquity and medieval kingdoms mustering their armies, to modern total war, overall population size has often proved

* If you found that such examples sprung easily to mind and concluded that the phenomenon of the smaller army prevailing isn't all that rare then your brain has succumbed to availability bias, one of the cognitive biases we'll discuss in Chapter 8.

decisive. The men marching into battle are only the point of the spear: it is an entire society that supports an army. Beyond the carpenters and cooks that moved with the soldiers on campaign, people at home made weapons and armour, bred and trained horses, constructed wagons or chariots, and worked in the fields to feed the army on the march. Of course, through much of history, the young, able-bodied men mustered into an army were the same needed to tend the fields at home. For millennia, therefore, there has been a seasonal pattern in the pulse of wars. Campaigns tended to avoid the harsher weather conditions of winter, and were usually launched in spring or autumn, so as to keep labourers in the fields for the most intense phases of the agricultural cycle, such as harvest.[32] In the twentieth century, the advent of total war saw mass conscription and the mobilisation of the entire civilian society for the struggle, which included women and older men working in armaments factories.

This dominance of demography in warfare has not escaped military thinkers and philosophers. The sixth-century BC Chinese general and strategist Sun Tzu advised in *The Art of War*: 'If your enemy is in superior strength, evade him.' In the first century AD, the Roman historian Tacitus was concerned by the small sizes of Roman families compared to those of the rapidly breeding German barbarian tribes beyond the borders.[33] An early-nineteenth-century proverb propounded, 'Providence is always on the side of the big battalions.'[34] And the Prussian general and military theorist Carl von Clausewitz wrote in his study *On War* (1833) that the superiority of numbers is 'the most general principle of victory'.[35]

The number of men bearing arms on the battlefield depends on the number of babies in the cradle two or three decades earlier.[36] And so the military might of a state rests on fundamental demographic factors such as birth rate. Unsurprisingly, therefore, states worry about their population growth and display mounting anxiety when the birth rate slows. This is what happened in France in the nineteenth century. For centuries, the country had supported the largest population in Europe, but that changed with the legacy left by Napoleon.

DEMOGRAPHIC EFFECTS OF NAPOLEON

Napoleon was undeniably a brilliant military tactician and accomplished statesman. Born into a lowly background, and riddled with insecurities, he was obsessed with his own prestige. He seized power from the leaders of the fledgling French revolutionary republic in a bloodless coup in 1799 and installed himself as the head of a more authoritarian regime. He made himself emperor in 1804 and set about reforming France's military, financial, legal and educational institutions. He left a vast legacy, not just within France but across Europe and around the world. Ultimately, it was his insatiable ambition and hubris that led to his downfall, which began with the ill-fated invasion of Russia in 1812. His Grande Armée, the greatest military force the world had seen, was swallowed by the freezing expanse of the Russian hinterland on the retreat from Moscow.[37]

Overall, around one million French soldiers died in the Napoleonic Wars of 1803–1815 and perhaps a further 600,000 civilians from the hunger and disease caused by conflict.[38] This represents almost 40 per cent of the conscription class of 1790–1795, and is a greater proportional loss even than that suffered a century later when France hurled its young men against imperial Germany in the First World War.[39]

A great number of young Frenchmen never came home, and this, coupled with the general disruption of war, caused a significant temporary drop in the birth rate. But long after the immediate effects of military conflict, throughout the nineteenth century, France continued to experience a very slow population growth, falling behind other European nations. This is partly due to the fact that France was slow to industrialise and remained a predominantly rural population with low living standards.[40] The decline in Roman Catholic observance after the revolution may also have contributed to smaller families.[41] But perhaps the main reason for the sluggish population growth was another legacy of Napoleon's rule.

Under the Ancien Régime, French law had been a patchwork of differing customs and rules across the provinces of the country; it was also riddled with special exemptions and privileges granted to the aristocracy by kings and lords. After the revolution, these last vestiges of the feudal system were abolished and fresh legislation was instated in accordance with the core tenets of the new republic: *liberté, égalité, fraternité*. When Napoleon seized power in 1799, he set about overhauling the French legal system from the ground up into a single, codified set of laws. The French Civil Code was crafted by a commission of jurists overseen by Napoleon himself and was enacted in March 1804. It became known, after its creator, as the Napoleonic Code.

Of particular concern to the post-revolutionary lawmakers was the overhauling of the old laws on inheritance that had enabled the dynastic transmission of wealth and power within feudal and aristocratic systems (as we explored in Chapter 2). These relics of the Ancien Régime were wholly incompatible with the ideals of the new republic.[42] Egalitarian principles for inheritance were already established prior to Napoleon. The National Assembly had decreed the abolition of primogeniture in 1791 and, by 1794, established the convention that the testator could only dispose 10 per cent of their property to a chosen heir, with the rest divided equally among all offspring regardless of sex or birth order.[43] But it was Napoleon who enshrined partible inheritance as articles within his codified civil law. And crucially, the Napoleonic Code also increased the disposable portion of inheritance the testator was free to choose to whom to bequeath it. Article 913 of the code specified that if a father had only one child, he could choose how to disperse half of the inheritance, with the rest going by law to the child; with two children he was allowed to decide what to do with one-third of the total legacy; and only one-quarter with three or more children.[44]

And this is where the demographic problem had its roots. While the motivation for crafting an egalitarian society was laudable, the inheritance provisions of the Napoleonic Code – the combination of enforced partible inheritance among all children along with a freely disposable portion based on the number of children – inadvertently created a strong incentive to have smaller families. With fewer children, the testator was allowed a greater disposable fraction of his total legacy and so was better able to prevent the dilution of family assets into the next generation.

France had for a long time been the most populous country in Europe. During the Middle Ages, more than a quarter of the continent's total population was French.[45] At the end of the eighteenth century, when Napoleon seized power, France still had the second-largest population in Europe (after Russia), with around 28 million people.[46] This was around 10 per cent more than the states that would come to be unified into Germany, and more than double that of Britain.[47]

Although, before 1800, marital fertility in France was similar to that of other European countries, the birth rate declined rapidly as the century wore on[48] – from around 30 live births per thousand people in 1800 to around only 20 by the middle of the century. By comparison, over the same period, the German birth rate fluctuated at around 37 births per thousand people, whereas in Britain it peaked at over 40 in 1820, before levelling off at around 36 through to 1880.[49] While the overall population of Europe more than doubled over the nineteenth century, France's grew by only 40 per cent. This happened despite France having one of the highest proportions of married women of any country in north-western Europe: French couples were just having far fewer children.[50] The fertility decline wasn't just among the richest classes, but across society. In France, most peasants owned their own land, in contrast to most of Europe. In the early nineteenth century, almost 63 per cent of the French population were landowning families (compared to only 14 per cent in Britain) and so stood to gain by limiting the number of children and thereby preventing the dispersal of their property.[51]

Just as France underwent a significant decline in its birth rate, another process was gathering pace in the rest of Europe, beginning with Britain.

For most of the history of civilisation, societies have faced high mortality rates, with deaths predominantly caused by malnutrition or disease. Consequently, in order to have a reasonable expectation that a least a few of their offspring would survive to adulthood, families needed to have a lot of children. These circumstances started to change in Britain from the mid-eighteenth century as the mortality rate began to decline because of greater food supply, improved public sanitation and advances in medical care. Mechanised agriculture and cheaper steam-powered transport of produce in particular improved food security. So as Britain industrialised, and its population became increasingly urban, the emerging gap between the birth and death rates produced a period of rapid population growth. This persisted until families responded to the lower mortalities by having fewer children, and the birth rate started to decline around 1880.

Britain was the first state in the world to undergo a modern, sustained population explosion, which gave it a significant advantage over its rivals on the world stage. In particular, this biological proliferation through the nineteenth century enabled Britain to export significant numbers of settlers to its overseas colonies and grow them rapidly without depopulating the homeland.* Industrialisation, trade and sea power were all

---

* Many of these settlers were economic migrants fleeing destitution at home. The nineteenth century saw huge population growth in the British Isles, but it was also a period of great social strife and disruption, with many in rural areas losing access to land or their livelihood, or, in the case of Ireland, facing devastating famine. Some of the dispossessed and desperate fled to the crowded, polluted industrial cities, but many set out for new lands overseas – sometimes with assisted immigration paying for their passage to the colonies.

important for the ascendancy of Britain into a global super-power, but the consolidation of the empire would not have been possible were it not for the demographic prowess of this small corner of Europe.[52] This is in stark contrast to the Spanish Empire between the sixteenth and early nineteenth centuries. As we have seen, the Spanish had laid claim to a vast expanse of territories around the world (see page 51), but there were simply not enough Spaniards to make a lasting population impact on the lands they had conquered even after they had succeeded – unintentionally or otherwise – in wiping out huge numbers of the indigenous inhabitants.[53]

This transformation from high death and birth rates, through a phase of lower death but high birth rates, and finally to low rates of both births and deaths, is known as the demographic transition. Much of the rest of Europe and North America followed Britain in undergoing this demographic transition, with industrialisation going hand in hand with a spurt of population growth before families began having fewer children. But uniquely in France, the fertility rate had already declined significantly from the beginning of the nineteenth century, and France only began to industrialise in the second half of the century.[54]

Unsurprisingly, the issue of France's low birth rate and relative underpopulation became a topic of public debate.[55] Many warned that as France's population diminished compared to its continental neighbours, it would become ever more vulnerable in the event of war, when more populous states would be able to field larger armies. Like Tacitus in the first century AD, the French were particularly concerned about the much higher birth rates of their eastern neighbours – the German states.

The growing German population came to surpass that of France by 1865, and five years later, France suffered a crushing defeat by Prussia, which led to the capture of Emperor Napoleon III, the occupation of Paris and the loss of the province of Alsace-Lorraine. And in January 1871, Bismarck united Prussia with a confederation of other German-speaking states, forging a new empire and creating the dominant land power in Europe. By the end of the nineteenth century, the languishing French population had plateaued at around 40 million. At the same time, the British population had quadrupled and now surpassed that of France, while the German population had surged to 56 million.[56] France had completely lost its superiority of numbers. Generations of slow population growth had placed France on the back foot. The French regarded their eastern neighbour with existential paranoia: there were fears that with the continuing high birth rate in Germany the next war would be even more catastrophic than the last.[57]*

In the years of mounting international tensions leading to the outbreak of the First World War, the European powers obsessively compared their relative industrial capability and population growth, like pugilists sizing each other up before a bar brawl. Britain and France both feared a growing Germany, while

---

* Differences in birth rate between separate sub-populations within the same nation or province can also be important. If a minority group sustains a greater birth rate and stands to become a more significant – or even predominant – fraction of the total population, it can disrupt the political status quo. The partitioning of Ireland in 1921, for example, demarcated the strongly nationalist Southern Ireland (which would become the Republic of Ireland in 1949) and Northern Ireland, composed of the six counties with a majority of unionists who desired to remain part of the United Kingdom. These unionists were on the whole the descendants of British colonists, and so generally Protestants, while there was a minority population of nationalists in Northern Ireland who were Catholic and wanted a united, independent Ireland. In the century since partitioning, however, the Catholic community has tended to have larger families and so maintained a higher population growth rate (although this has declined in recent decades), and by the most recent census in 2021, 46 per cent of the population reported they were brought up as Catholic, more than the Protestant community – a trend that has fuelled speculation that this demographic differential could swing the majority view in favour of nationalism.[58] In the May 2022 elections, Sinn Féin emerged, for the first time, as the largest party in the Northern Ireland assembly,[59] but it remains to be seen if this demographic trend and political shift will lead to a referendum that chooses Irish unification.

Germany in turn looked nervously at the rise of Russia. In the end, it was the international crisis triggered by the assassination of the Habsburg heir to the Austro-Hungarian throne, Archduke Franz Ferdinand (page 54), that lit the touchpaper. Germany reasoned that, since war with Russia seemed inevitable, it would be preferable to have that fight sooner rather than later – a gamble that activated a web of alliances and dragged Europe into war in July 1914.

With no significant technological disparity between the warring sides, Europe became locked in a strategic stalemate. Wave after wave of young men were sent into the meat-grinder of the trenches, and this conflict became the archetypal war of attrition. The sheer weight of numbers was paramount, with victory hanging in the balance for whoever could rally the larger population.

But the course and the outcome of the war did not depend on individual strength alone. While Germany's large population enabled it to field great armies in the Eastern and Western theatres, Britain could muster far fewer troops from its own soils. But the nineteenth-century mass emigration across the empire meant the British received the support of significant numbers of troops from Canada, Australia and New Zealand, as well as India and Africa. And while, with a population 40 per cent smaller than Germany's, France had a great deal to fear, it did not stand alone in 1914. Its alliance with Britain, Russia, Italy, Japan and, later, the United States ultimately swung the demographic balance. Even without the involvement of the US (whose expeditionary force didn't begin arriving in greater numbers on the Western Front until the summer of 1918), the Allies were able to mobilise almost 38 million troops; the Central Powers of Germany and Austria-Hungary, together with the Ottoman Empire, fielded fewer than 25 million. In the end, the First World War was won by the Allied Powers through their capacity to mobilise more troops for the battlefield. The cradles of the 1890s had proved decisive in the conflict.

DEMOGRAPHIC CONSEQUENCES OF WAR

Population size, and in particular the number of fighting-age men that can be called to arms, has been of paramount importance in wars throughout history. But warfare in turn also has profound effects on demography, with repercussions for society and the economy that endure for generations.

We have already seen how the destruction wrought by rampaging armies, plundering to supply themselves or otherwise disrupting the agricultural cycle, causing famine and spreading disease can cause widespread mortality across the societies impacted. In the last century, the Soviet Union suffered a particularly severe demographic catastrophe.

In June 1941, Hitler invaded the Soviet Union, starting what would become one of the deadliest conflicts in world history. Within the first six months, the Soviet Union had lost a huge expanse of territory – equivalent to roughly a third of the United States – and nearly five million men, as many as were in the Soviet Union's entire pre-war army.[60]

Critically short of soldiers, the Soviets significantly expanded the age range for conscription, including men under eighteen and over 55.[61] Overall, 34.5 million were drafted into the armed forces, nearly 8.7 million of whom died.[62] The total number of deaths, of soldiers and civilians combined, suffered by the Soviet Union during the Second World War is estimated to be around 26–27 million – around 13.5 per cent of the total pre-war population.[63] By comparison, the total number of casualties suffered by Germany amounted to 6–9 per cent of its peacetime population, and France and the UK each lost less than 2 per cent.

Alongside the immediate loss of life, the war's additional demographic impact was a crash in the number of babies born in the Soviet Union throughout the 1940s. Demobilisation and the return of the surviving soldiers to their homes took up to three years from the end of the war. The birth rate collapsed from 35 births per 1,000 people in 1940 to just 26 in 1946. This

25 per cent drop meant that an estimated 11.5 million babies were not born over these years due to the disruption of the war,[64] although the birth rate started to pick up again afterwards. The impact this had on the population structure in the Soviet Union was profound.

A common method for visualising a country's population structure is known as a population pyramid. A stack of horizontal bars represent the numbers of males and females in different age ranges, running from newborns at the bottom up to the elderly at the top. When a population is growing – when there are more babies being born than there are people dying – the society is youthful and the bar graph takes on the shape of a pyramid, hence the name 'population pyramid'.

The effects of both the cataclysmic number of deaths and the falling birth rate during the Second World War left an unmistakable imprint on the Russian population pyramid. The bars from the war years are markedly shorter, especially on the left side, where the male population is represented. In fact, the Russian population pyramid today shows a distinct pattern of shorter bars recurring roughly every 25 years – the approximate generation span. These are the demographic echoes of the devastation of the war. It shows that when the greatly reduced cohort of people born during and immediately after the war grew up and had children of their own, fewer children were born around 1968; and they in turn produced reduced numbers of offspring around 1999. The recurring demographic dent broadens each generation due to the spread in the ages at which people reproduce.

This wave-like structure of the Russian population pyramid – far more pronounced than any other nation impacted by WWII – has also affected Russian economics. Over time, as these waves progress upwards through the pyramid (as the population ages), the succession of peaks and troughs passes through the age window of the workforce – roughly those between 15 and 65 years old. A major factor in the economic strength of a nation is called the dependency ratio – the balance

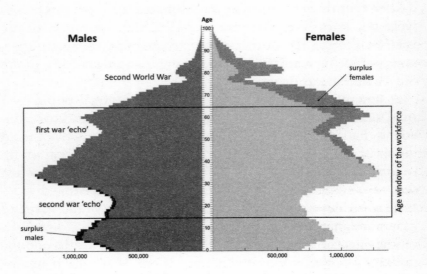

The population pyramid of Russia in 2022

between the fraction of the total population that is working and paying tax, and the number of dependents, the retired and children who are instead being supported. A high dependency ratio places a greater burden on the economy and reduces productivity growth. Because of the demographic reverberations of WWII, this dependency ratio has over the past 75 years fluctuated far more in Russia than elsewhere in the world, as either more peaks or more troughs in the population pyramid pass through the working-age window. Between the mid-1990s and late 2000s, the dependency ratio decreased significantly – the bulges in the population pyramid were moving through the working-age window – which is believed to have been a substantial contributing factor to the boom experienced by the Russian economy over this period. The World Bank estimated that nearly a full third of the growth in Russia's per capita GDP over this time was down to this demographic effect.[65]

But now that dependency tide has turned again, as instead more troughs pass through the working-age window, and the ratio is expected to increase significantly up to the early 2030s – more so than other nations experiencing a rising dependency ratio – placing increasing strain on the Russian economy.[66] So the particular features of the age distribution of the post-war Russian population continue to have a defining influence on the economy.

One particular subset of the demography is hardest hit by war – the young to middle-aged men predominantly marshalled to fight and die in the armies. Indeed, although the Second World War exacted a devastating toll on the Soviet population overall – killed by the Nazi occupiers or succumbing to the widespread malnourishment and disease – around 20 million of the 26–27 million deaths (75 per cent) were male, and most of these men were aged between 18 and 40.[67]

This removal of young men from the population severely distorts the sex ratio. Most animal species naturally have populations with an even balance of males and females. Although there might be an evolutionary advantage to an individual's genes by producing mostly sons – because, as we saw in Chapter 2, a male has the potential to produce far more offspring than a woman – in the next generation, now dominated by males, there would be a clear mating advantage to being one of the few females. So producing an equal number of males and females is known as an 'evolutionarily stable strategy'.[68] Human populations are also normally composed of a relatively even ratio of males to females.* Overall, at present, the global population is around 50.25 per cent male and 49.75 per cent female.[70] But on a regional level, extreme events can severely distort the 1:1 ratio – with long-term effects for society.

---

* Although, as women tend to live longer than men, in the ageing populations of developed nations today there is a slight female bias overall. In countries such as India or China, where there is a preference for sons, heightened by cultural practices or legislation such as the dowry system or a one-child policy, the overall populations are about 52 per cent male[69] (with the skew created by sex-selective abortion and even infanticide).

The jolt to the sex ratio in the Soviet Union from the Second World War was among the most extreme experienced by any country in the twentieth century. The sex ratio in the USSR prior to 1941 was already less than 1 due to the male death toll during the First World War, the revolution of 1917 and the civil war that followed. But the catastrophic losses during WWII skewed the ratio even further, especially in the western regions of the USSR where the fighting was most intense and the greatest casualties were sustained.[71] In the twenty years between the beginning of the war and the first post-war census in 1959, the number of draft-age men in the USSR had plummeted by more than 44 per cent, leaving 18.4 million more women than men in that age group.[72] This pushed the sex ratio down to as low as 0.64. The USSR was a country of women.

Many women had lost their husbands during the war, and with such a skewed sex ratio the prospects of finding a man to remarry, or for those never wed to find a partner, were dim. A long-term consequence of the scarcity of men, however, was a shift in male behaviour surrounding sex and marriage in the USSR. We've already explored humans' in-built predisposition to form strong-pair bonds for child-rearing, which formed the foundation of the cultural institution of marriage. But the reproductive strategies of men and women are not perfectly aligned. As men are not biologically bound to offspring in the same way the mother is, it pays for a woman to be choosy over potential mates. Her interests lie in selecting the best possible partner in terms of attributes such as genetic fitness, resources and social status, as well as commitment to co-rearing the child – a strategy that prioritises quality over quantity of mates. A male, on the other hand, has in theory no major limitation on the number of children he is able to sire and so may maximise his reproductive output by having children with many different women – if he can get away with it. This conflict between male and female reproductive strategies becomes all the more pronounced in a population with a grossly imbalanced sex ratio such as the post-war USSR.

In crude terms, the dynamics between male and female in mating or marriage are those of a marketplace, subject to the laws of abundance and scarcity, supply and demand. In a population where women are in short supply, they have greater bargaining power in the marriage market, and can take their pick of husband. And because the male is unlikely to find opportunities elsewhere, it's in his interests to remain committed to the relationship. In populations where there is a significant surplus of women, however, these priorities reverse. Women are in a weaker position to demand, and ultimately secure, male commitment and high levels of paternal investment in the rearing of offspring.[73] The scarce men have less incentive to commit to relationships and less fear of the consequences of seeking liaisons elsewhere, and there are plenty of other women willing to have a child out of wedlock.[74]

The missing generations of men in the post-war Soviet Union therefore drastically altered the market conditions. The sociological repercussions were clearly evident in the 1959 census: fewer women were married, and the divorce rates were significantly higher. The number of marriages with a large age gap between bride and groom had increased. And far more children were being born out of wedlock.*

The consequences of war-induced shortfalls of men were also evident in Germany after the Second World War. Although the fraction of men killed was much smaller than in the USSR, women in their prime fertility years aged between twenty and forty still outnumbered men in the same age bracket by a factor of ten to six in 1946. Here, too, many women struggled to secure husbands and remained unmarried.[75] Fertility rates dropped,

* The skewed sex ratio drove these changes in society, but the effects were likely magnified by the strongly pronatalist legislation of the Family Code brought in by the Politburo in 1944 to encourage births. State support was offered even for children born out of wedlock, creating a social environment that reduced the stigma of unmarried motherhood. Unmarried fathers were absolved of any legal or financial responsibility for their children – they didn't even need to be named on the birth certificate.

and the share of children born out of wedlock more than doubled to over 16 per cent. The southern German state of Bavaria,
with a predominantly Catholic population, recorded as many as
one in five babies born out of marriage, probably higher than
any other European region in the twentieth century.[76]

Thus the devastating loss of men in the USSR and Germany
in WWII produced a profound and persistent distortion in the
sex ratio; even today, the ratio in Russia is 0.87.[77] This demographic shock had not only significant consequences for the
birth rate but drove a shift in behaviour regarding sex and marriage and even changed attitudes towards gender roles. While in
Britain, the US and elsewhere, female empowerment and greater
financial independence brought by wartime mobilisation was
short-lived, as societies and economies returned to peacetime
norms and traditions,[78] the surplus of women in the USSR led
to long-lasting progressive cultural changes in attitudes towards
divorce, childbirth out of wedlock and premarital sex.[79]

STOLEN GENERATIONS

War is not the only human horror that can significantly distort
the sex ratio in populations.

From the early sixteenth through to the mid-nineteenth century, around 12.5 million Africans were captured and forcibly
transported across the Atlantic to the Americas to labour in the
plantations of European colonies.* Almost two million of them
did not survive the horrendous conditions of the oceanic voyage, and millions more would have perished in the raids and
wars in their homeland and during transport to the coastal factories where they were sold to European slave traders (numbers
that aren't included in the shipping records). There were other

* For a discussion on the transatlantic trade triangle, along with other oceanic trade routes that were significant during the Age of Sail, see Chapter 8 of
my previous book, *ORIGINS: How the Earth Shaped Human History*.

trade systems – the trans-Saharan, Red Sea and Indian Ocean slave trades – which, although less significant in scale and historical consequence, together exported a further 6 million people.[80]

The slave trades, an abhorrent stain on human history, have had profound long-term consequences for Africa. Across large areas of the continent, for hundreds of years Africans lived in fear that they or their families could at any point be seized into slavery. The removal of large numbers of people slowed population growth. Many factors can hamper population growth, including endemic diseases, crop failures and famine, but the demographic impact of the transatlantic slave trade – in terms of the staggering numbers of enslaved people, across a huge geographical region, over a long span of time – has no parallel in human history. The estimated population of sub-Saharan Africa in the early nineteenth century was about 50 million people; without the slave trades, historians have argued, it would have been 100 million.[81]*

Alongside the overall suppression of the population growth

* Historians have also identified a number of other long-term socio-economic repercussions of the transatlantic slave trade. Much of the continent is still hugely economically underdeveloped, even compared to other developing nations in South America and Asia, and there is a very strong correlation between the areas of Africa worst hit by slavery and those that are the poorest and least economically developed today.[82] In fact, over 70 per cent of the average income gap between modern Africa and the rest of the world has been calculated to be due to the destructive effects of the slave trade.[83] Part of this economic retardation may be due to the conditions of the slave trade driving a significant erosion of interpersonal trust, with neighbours and even kin frequently kidnapping or tricking one another – as many as 20 per cent of slaves may have been captured after being betrayed by a friend or family member. One of the long-running consequences of this culture of distrust has been a weaker foundation for the development of institutions and economic prosperity.[84] It's also been argued that this reinforced ethnic identities and impeded the formation of larger societies, which contributed to the fragmentation that we still witness today.[85] There appears to be a strong regional correlation between the intensity of the slave trade and the prevalence of internal conflicts in modern times.[86]

rate, the transatlantic slave trade also had a more specific impact. The demand driving the transatlantic slave trade was predominantly for the plantations in the American colonies. Plantation owners sought fit, strong labourers, and this meant that a peculiar characteristic of the transatlantic route – in contrast to, say, the Indian Ocean slave trade, which also sought female domestic maids and concubines – was a pronounced preference for male slaves. European slavers aimed at exporting twice as many men as women from Africa, and records show that the ratio of male to female slaves transported over the duration of the transatlantic slave trade stood around 1.8:1.[87]

Those regions most heavily targeted for slavery experienced a significant shortage of men.[88] At the peak of the transatlantic slave trade, at the end of the eighteenth century, there were fewer than 70 men per 100 women in West Africa. In Angola, the hardest hit area of the whole continent, the sex ratio dropped to as low as 0.5 or even 0.4.[89] Numerous communities found themselves with twice as many women as men.

Such severe distortions to the natural sex ratio drove a shift in family structure and altered the division of labour in society. Women had to substitute for the missing men and assumed the activities and responsibilities that had traditionally been the preserve of males in agriculture, commerce and even the military. They also adopted positions of leadership and authority in their communities.[90] This in turn changed the cultural norms and prevailing attitudes about the roles of women in society.

After the abolition of the transatlantic slave trade in the nineteenth century, the worst-affected areas began to return to a natural sex ratio of around 1, a balanced mix of male and female. But modified attitudes and practices around gender roles continued, and continue even to this day. With cultural norms passed down from parents to children, they persisted long after the disappearance of the demographic conditions that originally gave rise to them.

In the areas of Africa that endured the most intense burden of the slave trade, women today are more likely to be employed in the labour force, where they are also more likely to have a high-ranking occupation. Both men and women in these areas display more equal attitudes to gender roles, are less likely to accept domestic violence against women and are more supportive of gender equality in public office and politics.[91]

Importantly, these long-term social consequences of a skewed sex ratio are only found in the African regions most severely affected by the transatlantic slave trade: predominantly western Central Africa (especially Angola), sub-Saharan West Africa and, to a lesser extent, Mozambique and Madagascar on the south-eastern coast. Yet in East Africa, where the Indian Ocean slave trade captured more even numbers of men and women, there was less of an impact on the sex ratio, and cultural norms didn't change significantly; there's no evidence of increased female labour force participation, for example.[92]

Today, these regions also exhibit the greatest prevalence of polygyny.[93] It seems that the skewed sex ratios encouraged men to take multiple wives and women to accept the arrangement. And so the differences in demands serviced by the transatlantic and Indian Ocean slave trades may also explain why polygyny is much more prevalent in West Africa than in East Africa.[94] But the long-term repercussions of the demographic disturbance perhaps extend further.

Along with promiscuous forms of sexual behaviour, the practice of polygyny can drive the rapid spread of sexually transmitted infections such as HIV/AIDS, especially if the partners are unfaithful outside their relationships. There's not only a higher incidence of polygyny in West Africa but more infidelity by wives dissatisfied with their marriages. In this way, higher rates of HIV infection, particularly among women, across West Africa today may be driven by variations in sexual behaviour deriving from the grossly skewed sex ratios created by the transatlantic slave trade.[95, 96]

IT'S RAINING MEN

Circumstances that create demographic distortions in the opposite direction, to a high sex ratio with a surplus of men, are much less common, but there has been one prominent example.

In the decades leading up to the American Revolution, Britain was transporting around 2,000 convicts every year to its colonies in North America.[97] Overall, around 60,000 Britons were banished there[98] for crimes as trivial today as shoplifting and poaching or as bizarre as consorting with gypsies, using contraception and impersonating a Chelsea Pensioner.[99] But with the outbreak of war with the American colonies in 1776, this traffic came to an abrupt halt. British jails were soon filled to bursting and, judging it to be cheaper to transport convicts overseas rather than build more prisons, parliament became desperate for some other dumping ground for the nation's undesirables. The solution was to be found in a land on the other side of the world.

By 1770, Captain James Cook had charted the east coast of Australia and claimed half of the landmass for the British crown. Parliament now spotted an opportunity: not only could transportation to Australia serve as an emptying ground for the jails and a harsh deterrent to other criminals, but the convicts could form the vanguard of a new focus of colonisation in the southern hemisphere, supplying the labour needed to get the settlements established. In January 1788, the First Fleet arrived and established a settlement at Sydney Cove with around 1,500 settlers, 778 of whom were convicts.[100] The British colonisation of Australia had begun.

The numbers of convicts sent to Australia increased dramatically after the conclusion of the Napoleonic Wars, peaking in the 1830s when around a third of those found guilty in Britain's county courts[101] were sentenced to transportation and several years of indentured service. This practice of forced migration began tapering off in the following decade, however, with intensifying protests against the transportation system; and the last convict ship arrived in Australia in 1868.[102] By this time, over 157,000 convicts had been transported to Australia – mostly to the penal colonies established in New South Wales and the island of Tasmania – and 84 per cent of these were men.[103] Even when settlers began to come of their own free will from the 1830s, the number of men arriving in Australia still vastly exceeded women, not least because the available opportunities for work were mainly within arduous occupations like agriculture and mining.

The women living in Australia – whether emancipated ex-convicts (the sentence of indentured labour was generally seven years), free migrants or those born there – found themselves in an environment with a starkly skewed sex ratio. For a long period of Australian colonial history, the average ratio was about three men for every woman; it could be as high as 30:1 within some penal colonies.[104] Colonial Australia was, for almost 150 years, a land where it was continually raining men. The sex ratio didn't reach parity until around 1920.

Consequently, the value of each woman was of course much greater than in a balanced population, and a man would be exceedingly lucky to find a wife. It was very much a seller's market for the women. And this was significant for both sexes.*

Historical records show that during the colonial period, Australian women were much more likely to marry, and less likely to divorce, than their contemporaries in Europe. There is evidence that in populations with a surplus of males, the men are greatly incentivised to commit to any relationship they can

* This discussion is focussing on the colonists and their descendants in Australia. The First Nation peoples who had lived there for tens of thousands of years suffered greatly with the arrival of the colonists, who displaced them from their lands in often violent clashes and introduced new diseases; but the indigenous communities retained a roughly even gender balance.

secure without engaging in extramarital liaisons. They also dutifully participate in more childcare and make sure their wife is well provided for, so she has no cause to seek another husband. As a result, women in Australia felt no need to work and instead remained at home.[105]*

These male-biased conditions in colonial Australia have created enduring effects within society. What's astonishing is that even a century after the sex ratio balanced out to parity, the enhanced bargaining power of women is still imprinted on the expectations of gender roles and the statuses of men and women in society.

In areas of Australia today that historically experienced very biased sex ratios from the penal colonies, such as the region surrounding Sydney, along the north coast and on Tasmania, both men and women are more likely to hold conservative views on the role of women within society and to expect them to stay at home. Here, women still participate less in the labour market, work fewer hours, hold more part-time positions and are less likely to be in high-ranking occupations, compared to women in areas that were not so male-biased in the past. However, they do not spend more time on household chores or childcare instead; if anything, they spend less time on such activities. This means that women living today in areas that were once heavily male-biased enjoy considerably more leisure time over the week.[108]

Large-scale features of human populations – such as growth rate, population size and sex ratio – have had profound effects through history. These include the spread of agricultural communities, military might and war between states, and

---

* There is, however, a darker side to the patterns seen in populations with a large male surplus. High sex ratios are also linked to increased rates of sexual violence, predominantly committed by the single men left behind in the marriage market.[106] Sexual exploitation and abuse by white settlers, mostly against indigenous women, was prevalent in the colonial era.[107] Parallels could be drawn here to the most toxic aspects of the 'incel' counterculture today.

long-lasting social changes and impacts on economics. Another fundamental aspect of human biology is our predilection for consuming substances that alter our conscious experience. In the next chapter we'll see how, by changing our minds, such psychoactive substances came to change the world.

Chapter 6

Changing our Minds

If one would dispel an evening's unproductive lassitude, the meaning of 'drink' is tea ... Tea gives one vigor of body, contentment of mind, and determination of purpose
—Lu Yu, *The Classic of Tea*

Humans use plants for many purposes – for making food, for weaving clothes and for producing medicines. We also exploit the botanical world not only to enable our survival but to modify the functioning of our brain – to stimulate, to calm or to induce hallucinations. As conscious beings, we purposely consume substances with the sole intent of altering our state of mind. Indeed, enjoying getting out of our own minds is pretty much a universal of human cultures around the world.

Psychoactive drugs affect the functioning of the central nervous system – they alter our mood, our consciousness or our perception of the external world. We have learned how to self-administer various drugs to achieve different effects. In this chapter, we will examine four substances that change how the human brain functions and by doing so have changed the world: alcohol, caffeine, nicotine and opium. Alcohol and opium are depressants, whereas caffeine and nicotine are stimulants. Each has been used widely as a recreational drug – that is, one taken socially or for pleasure rather than for medicinal purposes. They affect our neurones in different ways, but they all trigger the reward centres of our brain, producing pleasure or elation. On the flip side of the cognitive coin, they are addictive and so drive further consumption.

This selection is not to imply that other plant products affecting our perception have not also been significant in certain cultures and societies. For example, for at least a thousand years, the indigenous peoples of the Amazon basin have been brewing ayahuasca to commune with the spirits of the natural world in social and shamanic rituals. Indigenous North Americans have consumed peyote, a small spineless cactus, as a source of the psychoactive drug mescaline for similar purposes for at least five thousand years. But these drugs did not become nearly as globally prevalent or influential as the four we are about to explore.

ALCOHOL

The most common method for altering our state of consciousness throughout human history has been the consumption of alcohol. Unlike particular plants with psychoactive properties, which are only available to societies living within its native range, alcohol can be brewed from a huge range of materials. In principle, any foodstuff containing sugars (such as fruit) or starch (such as grain or tubers), which can be broken down into sugars, can be fermented to produce alcohol.*

Fermentation is a general term for the use of microorganisms to change the properties of food and aid in the preservation of its nutrients; it is behind the production of yoghurt, cheese, soy sauce and pickled foods such as kimchi and kombucha.[1] But the earliest use of fermentation by humans was probably in the creation of alcohol.[2]

---

* Starch is a long-chain polymer composed of glucose molecules, which can be broken down into fermentable sugars by microorganisms, or even by chewing, with the enzymes in saliva doing the work.

Cultures around the world have developed their own local alcoholic beverages. Fermented sugar-rich substrates include wine from grape juice and cider from apples, while honey has been diluted with water and fermented to produce mead. Maple sap was fermented by the Iroquois in north-eastern America; the sweet pulp of cacao pods was used in Mesoamerica; and the sap of the agave cactus has been fermented in Mexico to form *pulque*. The starch-rich grains of cereal crops are also common starting materials and are used to produce Japanese rice wine or sake; maize beer or *chicha* in the Andes;[3] and *tiswin* in the south-western region of North America. India has long brewed beer from rice or millet. Cassava roots were fermented along the north-east coast of South America to produce *kasiri*.

Brewing has been a near-universal human pastime. Early evidence of wine-making is provided by residues – including the anthocyanin pigments which give red wine its ruby hue – found in clay jars dating back to around 3000 BC from Godin Tepe, an ancient Mesopotamian trading outpost in western Iran. Grape pips from plants in the early stages of domestication suggest wine production as early as 4000 BC in eastern Macedonia.[4] Wine became an important component of cultures in the Middle East before being adopted with bacchanalian passion by the ancient Greeks and then the Romans.

Some botanists even argue that ensuring a steady supply of fermentable grains was the driving motivation behind the earliest cultivation of cereal crops, suggesting farming was inspired by beer not bread.[5] This is an intriguing idea, but the archaeological evidence points to deliberate fermentation arriving after the domestication of cereal crops, indicating that alcohol was more likely a happy consequence, rather than cause, of the earliest grain cultivation and storage.[6] Nonetheless, brewing beer from grain certainly reaches back to the dawn of civilisation in Mesopotamia. And in a constitutional sense beer and bread are two sides of the same coin: beer is liquid bread, and bread is solid beer.[7] Beer-making spread from Mesopotamia to Egypt and then throughout Europe, being preferred in northern climes too cold for vineyards.

Alcohol has played many different roles in human societies. Drunk in moderation, its mind-altering effects lend themselves to celebration and joviality – alcohol reduces social anxiety and inhibition. It has also been instrumental in religious rituals, from ancient Egyptians offering wine to their deities, to Incans consuming large quantities of *chicha* to commune with the gods, to Christians performing the sacrament of the Eucharist.[8] But alcohol also served an important practical function: the presence of alcohol in fermented drinks such as beer and wine, or in distilled spirits added to water, kills many of the microorganisms that cause waterborne diseases.[9] A key step in beer brewing is to steep barley or another grain in water so that it germinates, and then boil this wort before fermentation, which kills any germs in the raw materials. From the Middle Ages until the nineteenth century, weak alcoholic beverages were routinely consumed in Europe, and then also in its colonies in North America, including by children, as a safe source of hydration since rivers and wells were usually contaminated.* (Other cultures took to drinking infusions of boiled water – such as tea in China.)

EFFECTS OF ALCOHOL ON THE BRAIN

Alcohol is absorbed readily into the bloodstream, and within minutes, it binds to receptors in the brain to suppress the release of excitatory neurotransmitters and increase the effects of inhibitory ones. The overall effect of alcohol is therefore as a sedative. This may surprise you if you feel that a drink livens you up and makes you feel more social. But this is because even after your first drink the depressant effects of alcohol are already

---

* It is perhaps no coincidence, therefore, that temperance and prohibition movements only arose in the late nineteenth century, after governments had begun large projects to supply urban populations with clean drinking water.[10]

starting to take hold in regions of your brain such as the pre-frontal cortex, which is responsible for higher functions such as moderating our social behaviour and constraining impulses. As this brain region becomes sedated, this control lessens and we start to loosen up, feeling less anxious or self-conscious and more extroverted. Thus small amounts of alcohol can elevate your mood and help you relax, acting as an indirect stimulant. If you continue to drink, however, more and more of your brain becomes stupefied. Higher doses of alcohol have a dangerously debilitating effect on your senses, motor control and cognition, as they poison the brain and cause blurred vision, loss of balance, slurred speech, confusion, blackouts, memory loss and nausea. Extreme levels can lead to loss of consciousness, a coma or even death.

The body breaks down ethanol using a group of enzymes called alcohol dehydrogenases. The dehydrogenase enzyme that is found in humans and our great ape cousins disposes of ethanol around forty times faster than those also found in other species. This fast-acting version seems to have evolved around 10 million years ago. This far predates the deliberate brewing of plant material, of course, and probably arose at the time when our ancestors came down from the trees and started spending more time on the ground, eating fallen fruit that had started to ferment naturally.[11]

We also possess a number of other enzymes that work on the break-down products of ethanol, such as the toxic compound acetaldehyde, the build-up of which is one of the causes of the hangover. Some variants of these produce an unpleasant reaction. For example, one gene variant common in East Asian populations leads to a flushed, red face, nausea and headaches after a drink, affecting over a third of people in China, Japan and Korea today.[12] Historically, these cultures have consumed far less alcohol and their populations today are much less affected by alcohol abuse.[13] These genetic differences were almost certainly driven by differing cultural practices around the world over the last 10,000 years, with variants of the enzyme that were more effective at scrubbing alcohol from the bloodstream selected for in those populations that drank more often.

For reasons we'll return to later, the psychotropic effects of alcohol are addictive. Both this addiction and the behaviour-altering effects of alcohol mean that its consumption does not only have a direct impact on the drinker but can harm those around them as well. The social ills caused by alcohol abuse are legion: alcohol causes more accidents, injuries and violence than any other drug. Smoking is also highly addictive and harmful, but no one starts a pub brawl or causes a motorway pile-up because they had one too many cigarettes.[14]*

DISTILLATION

The effects of alcohol are more potent in more concentrated alcoholic beverages. During brewing, yeast cells metabolise sugar into ethanol, proliferating until the alcohol content has increased to the point that it prevents any further growth and they start to die off, essentially poisoned by their own waste products. This happens at an alcohol concentration of around 14 per cent, which is the maximum booziness that fermented beverages can attain. To achieve a higher alcohol content a new

---

* Governmental efforts to control the menace of alcohol can have unintended consequences. In the United States, the temperance movement led to nationwide prohibition in 1920 with the Eighteenth Amendment to the US Constitution. Outlawing the manufacture, transportation and sale of intoxicating liquors, however, did nothing to remove the demand for alcohol in society: illegal stills and speakeasies proliferated rapidly, and within just a few years, alcohol consumption had risen to around two-thirds of pre-Prohibition levels.[15] The draconian law only encouraged the black market, bootlegging and the growth of organised crime until the ill-conceived policy was abandoned. The Twenty-first Amendment to the constitution in 1933 was the federal repeal of the nationwide prohibition on alcohol.

technology is needed – distillation. This process exploits the fact that, at 78 °C, ethanol has a lower boiling point than water. So, if you heat a mixture of water and ethanol produced from fermentation, the ethanol vapour comes off first and can be cooled to condense back into a concentrated liquid. Distillation was used thousands of years ago in China and the Middle East to extract fragrant oils from rose petals to create perfumes or produce medicines as well as spirits. Simple stills such as the alembic, using a single side-arm to collect and cool the ethanol vapours, weren't particularly efficient, but more complex apparatus, such as those equipped with a coil of copper pipe cooled in a reservoir of water, and accurate temperature control of the distilling vessel, can produce concentrated ethanol, especially if the distillation is repeated. If the distillate exceeds about 50 per cent ethanol, it can be set alight and is said to be 100 'proof'. The distillation of wine, for example, yields a potent alcoholic drink known in Dutch as *brandewijn* ('burnt wine'), which became 'brandewine' in English, or simply 'brandy'.

Not only do distilled spirits offer a concentrated form of alcohol, but they also don't spoil like beer or wine and so can be transported long distances.* This makes spirits useful and high-value trade items. For indigenous American cultures, for example, which were already familiar with intoxicating substances such as tobacco and peyote, alcohol, and especially distilled spirits, became an important component of the gift exchange and then trade with the first European colonisers.[17]

Distilled drinks also played a central role in the transatlantic slave trade. The African slavers who supplied the Europeans with captives valued a range of commodities in exchange,

---

* A method for safeguarding beer was developed in the early 1800s for provisioning British troops in India. The voyage to India took six months or more, much of it through the hot tropics, and beer in the holds of East Indiamen often arrived spoiled. London brewers started to add fresh hops to act as a preservative, and so the distinctively flavoured India pale ale, or IPA, was born.[16]

including textiles, metals and, especially, distilled spirits such as brandy, which they preferred to their own locally produced, and less concentrated, grain-based beers and wines. The link between spirits and slavery was only strengthened by the invention of a potent new distilled drink made from sugar cane. Rum had been made from cane juice in Brazil by the Portuguese from the early sixteenth century, but the British refined the process in Barbados in the mid-seventeenth century to use molasses, an otherwise worthless waste product of sugar production. Not only was rum therefore cheap, but as a distilled spirit it offered a compact, self-preserving form of alcohol and so became a key part of the transatlantic economy. Rum could be used to buy slaves, who laboured to cultivate sugar, the leftovers of which could be converted into rum to buy yet more slaves.

Distilled spirits also became a key enabler of longer sea voyages. Fresh water stored in barrels on board ships quickly went stale and developed microbial growth, and so this stagnant water was often sweetened with beer or wine to make it palatable. But these too spoiled on longer voyages. When concentrated spirits became affordable and widely available, they were used instead to temper the drinking water. After 1655, the English replaced a sailor's daily ration of beer with rum, first in the Caribbean and then for all Royal Navy ships. Yet once sailors had developed a habit of saving up several rations to drink all at once to become intoxicated, in 1740, Vice-Admiral Edward Vernon ordered that the standard ration of half a pint of rum be diluted with four parts of water, thereby creating grog which was issued half at the noon bell and half at the end of the day.*

---

* The drink acquired its name from Vernon's nickname 'Old Grog', on account of his predilection for coats made of a coarse woollen material called grogram.[18] The practice of serving grog twice a day was later also adopted by the American Continental Navy.

DOPAMINE AND THE BRAIN'S PLEASURE CENTRE

At the top of the brainstem – one of the most ancient parts of our brain and the critical connective for the spinal cord – sits a cluster of neurones known as the ventral tegmentum.[19] The ventral tegmentum communicates with the nucleus accumbens, a behaviour-mediating region of the brain, via a tract of dopamine-releasing neurones called the mesolimbic pathway, and although they constitute only a minuscule portion of all the nerve cells in the brain – less than 0.001 per cent – these neurones are enormously important in motivating our behaviour towards survival and reproduction.[20] Eating food, quenching thirst or having sex all result in the release of dopamine in the mesolimbic pathway. Watching porn or even thinking about sex is enough to trigger dopamine release.[21] The dopamine system is also activated by other gratifying experiences, including getting revenge – such as we explored in Chapter 1 – or winning at a computer game.[22]

The reward signal is perceived as the sensation of pleasure, and so dopamine is often described as the pleasure chemical of the brain. And this dopamine-release mechanism doesn't just operate in humans. The mesolimbic reward pathway is shared by all mammals – it is an ancient, fundamental part of the brain's functioning – and indeed, similar systems involving dopamine or related neurotransmitters for directing behaviour are universal across the entire animal kingdom.[23]

The mesolimbic pathway flushes with dopamine in response to a favourable outcome, such as eating or drinking, and especially one that was unexpected; conversely, dopamine levels are decreased by a negative experience or an anticipated pay-off that doesn't materialise. So in order to tune our behaviour to succeed in our natural habitat, our brain compels us to repeat the actions that activated the dopamine system last time and avoid those that previously suppressed it. Thus the neurochemical system of pleasure and reward is inextricably intertwined with that of learning. The dopamine pathway also connects the tegmentum to the prefrontal cortex, the wrinkly part right at the front of the brain that is greatly enlarged in humans compared to other animals. The prefrontal cortex orchestrates high-level 'executive' functions such as making decisions and planning towards particular goals and so is ultimately informed by the dopamine reward pathway.

This dopamine-mediated mechanism works exceedingly well at steering our behaviour towards the sort of actions that benefit us in the natural world. Problems arose, however, when humans discovered ways of triggering this reward-and-pleasure system with stimuli other than those associated with increasing biological fitness – namely, drugs.

Alcohol, caffeine, nicotine and opium effectively short-circuit our brain's reward system. They induce the release of dopamine in the mesolimbic pathway – or inhibit the removal of dopamine or make the receptors on the surface of neurones more sensitive – and in some cases they can produce pleasure, even euphoria, far more intense than anything encountered in the natural world. And unlike natural dopamine triggers, such as eating, they never result in satiation.

Drugs create a false signal in the mesolimbic pathway that indicates the arrival of a huge survival benefit, and the learning mechanism driven by this system rewires the brain to seek out repeated hits. This is the basis of addiction, and it produces cravings and compulsive behaviour to achieve that instant gratification without having to pay the costs associated with dopamine rewards in the natural world, such as spending time hunting for food.

Experiments conducted in the 1950s surgically implanted wire electrodes deep into the brains of lab rats so that the nucleus accumbens would be stimulated every time they pushed a lever. They found that the rats would compulsively do this – up to 2,000 times an hour – foregoing drinking, eating, sleeping or any other normal behaviour to keep triggering buzzes of pure pleasure, until they collapsed with exhaustion.[24] The sad truth is that humans can become caught in similar traps, administered not by

electrodes directly stimulating the brain but by chemical substances similarly targeting the mesolimbic reward pathway. The
problem is worse when the natural plant product has been
refined and the psychoactive compound concentrated, or even
chemically modified, to increase its potency – heroin synthesised
from raw opium, for example. Delivering the active compound
in a more rapid surge to the brain also greatly increases its
euphoric effect and addictiveness, such as by smoking, snorting
or especially injecting directly into the bloodstream, rather than
ingesting orally.

The dopamine system also recalibrates itself, so that with
repeated large rewards the dopamine release is dialled down
back to its base level. This is the process of habituation and the
reason junkies – whether they've developed a habit for coffee or
cocaine – need ever-greater doses to experience the same buzz.
As the neuroendocrinologist Robert Sapolsky puts it, 'What
was an unexpected pleasure yesterday is what we feel entitled to
today, and what won't be enough tomorrow.'[25] Before long, the
pleasurable sensations that the drug once produced have faded
away, and continued use becomes motivated instead by trying
to avoid the nasty effects of withdrawal. In this way, drugs very
effectively hack our brain's reward system for tuning behaviour
for survival, and so substance abuse is a universal human
vulnerability.

Alcohol has been the most widespread method for altering
our state of consciousness through human history, and its ability to light up the dopamine pathway yields the warm glow of
inebriation as well as the development of dependence. After
alcohol, caffeine has been humanity's second-most popular
drug.

CAFFEINE

Caffeine is the most widely used psychoactive stimulant in the
world, and coffee is one of the most valuable commodities

exported by developing nations.[26] (Alcohol has been more widely consumed by cultures throughout history but, as we have seen, is a depressant rather than a stimulant.) Around 90 per cent of the world's population consume the drug regularly in some form or other, including children, given the caffeine in soft drinks. Besides tea and coffee, a few other plants also produce caffeine, such as cacao (chocolate), kola, guarana, yerba mate and yaupon – all of which have been infused in water to create caffeinated drinks by people living within their botanical range.[27] Today, tea and coffee are consumed enthusiastically in all seven continents – including by the research scientists stationed on Antarctica – and there is even an espresso machine aboard the International Space Station called, punningly enough, the ISSpresso maker.[28]

The apocryphal story on the origins of coffee consumption goes something like this. A ninth-century goat herder in Ethiopia noticed that his goats became frisky and uncontrollable after eating the cherry-red berries of a particular bush. He then tried some himself and was pleasantly surprised by their invigorating effect. In time, processing developed, and the seeds were roasted and then ground into a powder to be infused in boiling water – and the practice of making a cup of coffee was born. But whoever discovered coffee, the Sufi mystics in Yemen, just over the Red Sea from Ethiopia, were early adopters of the drink.[29] These Muslims customarily prayed late into the night, and coffee helped them stay awake and alert into the early hours. It was the invigorating effects of coffee that kept the whirling dervishes spinning in their religious rituals. The port of Mocha (or Al-Makha) in Yemen became a major trading hub for coffee from the Horn of Africa and provided one of the names for coffee (today often mixed with milk and chocolate). Coffee reached Constantinople in the sixteenth century and quickly spread throughout the Ottoman Empire and around the Mediterranean. Indeed, it is the Turkish word *kahve*, morphing into 'caffe' in Italian, that gave us our name for this compelling drink. The embracing of coffee by the Islamic world owed as much to its energising effects as to the fact that, unlike alcohol, it was not generally interpreted as being forbidden by the Qur'an – it became known as the 'wine of Araby'.[30]*

Coffee seems to have arrived in the maritime trading hub of Venice in 1575, and by the mid-seventeenth century, it was readily available across Northern Europe. The first coffee house in London opened around 1652, and within a few decades there were thousands in the capital alone. This rapid growth was certainly driven by caffeine's addictiveness: once a customer had drunk a cup and felt the effects, they were likely to return again and again. But just as important for the rise of coffee was the culture that came to surround its consumption. Unlike the alcoholic beverages drunk in taverns that dulled your senses and made you sleepy, coffee revived and energised its drinkers. Coffee houses therefore emerged as places not only for friends to relax and catch up in, but also for businessmen of the new mercantile class to negotiate deals, and for intellectuals to debate ideas at the dawn of the Enlightenment. They were a new kind of public space, great democratic levellers where men from across the class divide could congregate alongside each other and pick up an education by listening to surrounding conversations – they became known as 'penny universities'. Coffee houses also came to provide newspapers and printed pamphlets, drawing customers in to catch up on the latest news and ideas and share rumours with their fellow drinkers. With coffee houses becoming hotbeds of debate, free thinking and political dissent, Charles II attempted to close them in 1675, especially because the main topic of chatter was the king himself and the fate of the Stuart restoration.[32] The coffee houses of

---

* Coffee, of course, was only discovered after the time of Mohammed and the writing of the Qur'an, and so it couldn't have been explicitly prohibited. Similarly, the Word of Wisdom of the Church of Latter Day Saints forbids the consumption of alcohol, tobacco and caffeinated drinks, which all existed when the movement was founded, but today, young Mormons will happily partake of MDMA (ecstasy) at raves.[31]

Paris too were alive with political discourse, even sedition, and played an instrumental role in the events of 1789. The Café Procope was frequented by Robespierre and other prominent revolutionaries, and it was from a table outside the Café de Foy that Camille Desmoulins roused the mob that stormed the Bastille – the insurgents fired up not on alcohol but on caffeine.

Particular coffee houses became closely associated with different areas of business, and patrons knew they would lose a competitive edge over their rivals unless they frequented the right establishment to keep abreast of developments. Lloyd's of London, for example, began as a coffee house popular among merchants and shipping magnates and so was a crucial hub for the latest news on the coming and going of ships and their cargoes. It became a major centre for obtaining maritime insurance, and by the late nineteenth century, the corporation was one of the largest underwriters in the country. Likewise, the London Stock Exchange began in a coffee house. Coffee therefore lay right at the heart of not only intellectual development and the Enlightenment but capitalism and the birth of many of today's financial institutions.

Our passion for tea and coffee was also a major force in driving long-distance maritime trade and shaping the global economy. Until the early 1700s, all coffee reaching Europe was cultivated in Yemen and came from Mocha. But the Dutch East India company (VOC) started planting the crop – descended from plants originally smuggled out of Yemen – in its colonies in the East Indies in the 1690s, and by the 1720s, Amsterdam had become the coffee capital of the world, with about 90 per cent of the beans moving through its exchanges originating from the Dutch-owned island of Java. Other European imperial powers took to supplying their domestic demand by growing coffee in their colonies too, most of it on the back of slave labour. The French took it to Martinique and Saint-Domingue, and by the 1770s, well over half of the world's coffee was produced in Haiti. The slave revolt at the end of the century destroyed many of these plantations, and the refusal of the European states to recognise or deal with this first black republic ended this lucrative trade (as we explored in Chapter 3).

The Portuguese colonies in South America used slash-and-burn cultivation to grow the coveted crop, and when the soil became depleted, they moved on to another tract of land, resulting in the deforestation of the Atlantic seaboard of Brazil. These huge plantations and the slave labour – Brazilian coffee plantations used slaves as late as 1888[33] – were able to produce large quantities of high-quality coffee at low prices, supplying the growth in demand for coffee in America throughout the nineteenth century. Brazilian coffee exports leaped 75-fold between the country's independence in 1822 and the end of the century, when it was producing almost five times as much as the rest of the world combined. It was this colossal Brazilian supply that led to plummeting prices and made a cup of coffee the mass-market commodity it is today.[34, 35]

Tea-drinking has had an even longer history than coffee consumption. Tea was originally used as a medicinal concoction in the Yunnan province of south-west China by about 1000 BC. But green tea became broadly popular as a stimulating hot drink across the whole of China, and then South East Asia, from the mid-eighth century AD during the Tang dynasty.[36] As with the Sufis, it was adopted early by Buddhist monks striving to stay awake and focus during their long meditations.[37]

Tea leaves were first brought to Europe by the VOC in the early seventeenth century, and the beverage began appearing in English coffee houses in the 1650s. This international trade to the west favoured fermented and oxidised black tea, which kept better than green tea, and since the VOC dominated the coffee trade, the British East India Company came to focus its efforts on the tea grown in China.

Through the second half of the seventeenth century, the expense of tea in Britain meant it was largely consumed for its supposed medicinal properties or as an aristocratic status symbol. By the mid-eighteenth century, however, the British East

India Company was importing so much tea from China that prices fell enough for tea drinking to become a domestic ritual for the middle classes, and soon it was widely consumed by the working classes too. Tea was now the national hot beverage, sipped and slurped right across society, from the royal palaces to the dwellings of rural labourers and the urban poor.[38] British drinkers also took to another innovation, that of adding milk and sugar to their cup of tea – a practice unknown in China. This new habit contributed to the expansion of slavery in the sugar plantations of the Caribbean, as it further increased demand for the sweet stuff. The desire to break the Chinese monopoly on the invigorating leaves drove the East India Company to begin large-scale cultivation of the native tea plants in the north-east Indian region of Assam in the early nineteenth century. Then, in an early historical example of corporate espionage in 1850, Scottish botanist Robert Fortune smuggled tea plants out of China, and large plantations were established across Darjeeling.

In America, the story of tea took a very different trajectory. It was introduced to the Thirteen Colonies at around the same time as England and similarly grew in popularity. The East India Company endeavoured to supply this demand, but by the late 1760s, most of the leaves consumed in the colonies were smuggled Dutch tea. Indeed, the drinking of smuggled tea was encouraged by American patriots such as the Sons of Liberty as a political protest against British taxation. The East India Company saw surpluses of tea piling up in its London warehouses that it couldn't sell. In an effort to prop up the financially beleaguered company by helping them to undercut the price of smuggled tea, the British parliament passed the Tea Act in 1773. This allowed the company to ship tea directly from China to America, without having to pay British import duty, and granted them a monopoly on the sale of tea in America. The tea was subject to tax only in the colonies.

The colonists, however, viewed this as an attempt to foist British taxes on them. They harassed East India Company consignees, refused to accept the tea and left it to rot on the dockside, or prevented the East Indiamen from landing the product. One of the most public – and famous – displays of rebellion flared in Boston Harbour, where in December 1773 protestors boarded the ships and destroyed over 340 chests full of tea by dumping them over the sides into the harbour waters. This Boston Tea Party triggered similar acts of rebellion in other ports, including New York. The situation escalated with parliament in 1774 passing the Coercive Acts – or the Intolerable Acts, as they were known in America – designed to make an example by punishing the defiance of Massachusetts, stripping the colony of self-governance and forcing the closure of Boston Harbour until the ruined cargo had been paid for. But these harsh reprisals only served to unite the colonies against the king, and tensions continued to mount until the War of Independence erupted the following spring.

The passion for coffee rather than tea in America today was not driven by the Tea Act and protests against British tea before the Revolutionary War. Tea remained popular, although it is perhaps telling that the Declaration of Independence was first read publicly outside the Merchant's Coffee House in Philadelphia.[39] But coffee grew in popularity in the decades after independence – it could be imported by the Americans directly from French and Dutch colonies in the Caribbean[40] – and in particular after 1832 when import duty on coffee was scrapped, making it cheaper.[41]

CAFFEINE'S EFFECT ON THE BRAIN

Caffeine is a molecular mimic. Every minute we are awake, a chemical called adenosine accumulates in the brain, marking the time since we awoke like sand in a sand timer. It gradually slows our mental processes and prepares us for sleep – creating what is evocatively referred to as a mounting sleep pressure[42] – so that after twelve to sixteen hours of being awake we once again feel the irresistible urge to lay back down to slumber.[43]

The particular shape of the caffeine molecule, however, just so happens to allow it to fit snugly into the same receptors that adenosine binds to, but it doesn't trigger them; instead it effectively maintains a chemical blockade of the adenosine ports. So, if your brain is flushed with caffeine, adenosine can't get to the receptors and the normal signalling is muted. Caffeine pharmacologically holds sleepiness at bay and keeps our brain in an alert, focussed state. The adenosine is still there, accumulating in the brain, but its signals are jammed by the caffeine. As the body breaks down the caffeine, the adenosine that has been building up behind the dam breaks through and releases an overwhelming wave of doziness – the dreaded caffeine crash.[44]

Caffeine is synthesised by plants as a natural pesticide, deterring insects from feeding on their leaves or seeds or even killing them.[45] But curiously, several kinds of plant, including those of the coffee and citrus subdivision, also produce caffeine in their nectar – the sugar-rich fluid secreted by flowers to *attract* insects for pollination. Experiments have shown that caffeine enhances olfactory learning in honeybees – they are more likely to remember the scent of the flower associated with the nectar reward and thus keep returning to caffeinated flowers. It's as if the plants are using the stimulant drug to hack the brains of the bees to recruit them as dedicated pollinators for their flowers; you could say that caffeine gives the bees their buzz.[46]

Another effect of caffeine is that it causes increased levels of dopamine in the nucleus accumbens, while it also heightens the sensitivity of dopamine receptors. This triggers the mesolimbic reward pathway we explored earlier, producing the buoyant, mood-enhancing properties of a good cup of tea or coffee but also its addictiveness.[47] Humans adopted drinks like coffee and tea because of their effects on our brains, serving as stimulants and suppressing sleepiness, and, once started, the compulsion of caffeine addiction maintained the habit. And in turn, caffeine has had a lasting effect on history.

While coffee stimulated the minds and discourse of European

intellectuals in coffee houses during the Age of Enlightenment, it was tea that enabled the bodies and minds of the English working classes to adapt to the changing industrial world. The Industrial Revolution swept aside traditional, specialised crafts such as weaving and blacksmithing and replaced them with heaving machinery. Artificial light, first gas lamps and then electric bulbs, enabled factories to operate late into the night. Caffeine not only kept the workers' brains alert and attentive through the monotonous factory work but also suppressed the hunger pangs of the undernourished. The sugar in the tea provided calories to sustain people's bodies through the long shifts. Caffeine transformed human workers into better accessories to the untiring iron machines they were serving.*

So, while coal powered the steam engines of the Industrial Revolution's factories and mills, tea supplied by the East India Company, sweetened with sugar from the West Indies, fuelled the workers tending the machines.[49] In this way, the history of tea is rooted in the exploitation of labour – from where it was grown in India to the sugar plantations in the Caribbean to the factories in Britain that squeezed every waking hour out of their workers.[50]

Today, caffeine plays a central role in controlling our natural sleep-wake cycles. Our fast-moving technological society no

---

* For similar reasons, psychoactive drugs have also been used in warfare. The astonishing speed of Hitler's blitzkrieg advance across Poland in September 1939 and then through France and Belgium in early 1940 was based on the mobility of the Wehrmacht's Panzer divisions, with the tanks equipped with radio sets for coordination and supported from the air by Luftwaffe bombers. But their success was also enabled by another technology. The synthetic stimulant drug methamphetamine (which has a similar molecular structure to the hormone adrenaline) allowed troops to fight harder and longer without feeling mental or physical fatigue. Meth chemically created a state of hyper-alertness while also boosting confidence and aggression. The lightning swiftness of the blitzkrieg drives owed their success to the fact that the troops were on speed. The Führer himself also took cocktails of injected drugs – including cocaine, methamphetamines and testosterone – to fuel himself through the war.[48]

longer allows us to respond passively to our biological rhythms; we are expected to adapt them to the dictates of its digital clock. Many of us self-administer caffeine to help meet these demands – when rousing ourselves for the daily commute, pulling an all-nighter at the desk or re-synching our body's rhythms to a new time zone after a long-haul flight. Many caffeine addicts are able to manage their own dosing of the drug to masterfully harness its positive effects and improve performance in the focus-demanding tasks of the modern world, while avoiding the negative consequences of overconsumption such as jitteriness, a racing heart or stomach irritation. But by enabling us to inhibit our brain's own signalling mechanism for sleepiness, caffeine is one of the major causes behind the modern pandemic of sleep deprivation. And what's so insidious about our relationship with coffee and tea is that the very thing we use to remedy the resultant chronic drowsiness is again caffeine.[51] In fact, much of the prompt to reach for the first cup of coffee or tea in the morning – to clear a foggy mind or help us wake up – is to relieve the symptoms of drug withdrawal after an overnight absence.

NICOTINE

As we saw in Chapter 4, the establishment of contact between the Old and New Worlds in the late fifteenth century irreversibly changed both. The diseases the European explorers brought with them utterly devastated the native populations of the Americas. In return, syphilis made its way across the Atlantic in the opposite direction, sailing back with Columbus's crew after that first voyage. But the global shift in the Earth's biota that ensued over the following decades and centuries, known as the Columbian Exchange, also brought back to the Old World something far more debilitating and deadly than the syphilis bacterium.

Today, tobacco use kills over 8 million people every year: about 15 per cent of all deaths worldwide result from the litany of cancers, cardiovascular and respiratory diseases it causes. Or to put it into perspective, it's as if the 1918 influenza pandemic was occurring every decade. And it's not just the actual users who are affected, but their children or anyone else breathing in their smoke.[52] In short, tobacco is the largest preventable cause of death in the world today.[53]

Tobacco belongs to the Solanaceae family of plants, making it a close relative of the deadly nightshade, which also includes the potato and the aubergine. The botanical genus of tobacco, *Nicotiana*, contains around 70 different species of plant,[54] but two have been most widely used by humans: *Nicotiana rustica* and *Nicotiana tabacum*.

Discoveries of nicotine residue in pipes show that tobacco smoking was practised in pre-agricultural hunter-gatherer societies in south-eastern North America over 3,000 years ago; and its use in South America can be dated to around the same time.[55] But humanity's relationship with the plant may be far more ancient. A recent archaeological find of charred tobacco seeds in a hunter-gatherer hearth in Utah suggests the use of tobacco as early as 12,300 years ago. This is not long after humans first arrived in the area after migrating into North America along the Bering land bridge during the last ice age.[56] The earliest cultivation of tobacco started in the Peruvian Andes and spread as far north as the Mississippi valley by around 500 AD.[57] Some North American tribes, such as the Blackfoot and the Crow, practised no other form of agriculture save planting and tending to their tobacco plants, showing just how central it was to their culture, while they gathered everything else they needed from the wild.

For the peoples of both North and South America, tobacco was a spiritual herb, consumed in highly ritualised ways. North American tribes smoked tobacco in long pipes as part of sacred ceremonies or to seal a treaty, the rising smoke serving as an offering to the gods. To inhale smoke was to receive into your body the ethereal; and on exhaling, your questions and desires were carried forth in a form fit for the spirits.[58] A shaman might bestow a blessing or protection on a warrior before battle by ritualised smoke blowing. In the South American civilisations, too, tobacco held profound religious importance. A Mayan manuscript referred to as the Madrid Codex, written in the fourteenth or fifteenth century, before much of the Mayan culture was obliterated by the Spanish conquest, depicts three deities smoking cigars.[59]

This connection between tobacco smoking – or sometimes chewing – and the spiritual can perhaps be explained by the fact that the species of tobacco plant used traditionally in the Americas, *Nicotiana rustica*, is much stronger than other varieties, such as that grown commercially today. Its leaves contain five to ten times the concentration of nicotine and can produce a narcotic effect like drunkenness or even bring on a trance and hallucinations. Shamans or priests would use high doses to commune with the spirits or induce visions.[60]

Tobacco was used for more practical applications too. It was known to alleviate thirst and hunger[61] (predating supermodels smoking to suppress their appetite).[62] Tobacco was also believed to have medicinal properties – somewhat ironic to us in the modern world now aware of the wide range of health problems it causes. It was taken to relieve or cure conditions such as asthma, toothache, earache, digestive problems, fever and depression, and was spread as a poultice onto wounds, insect bites or burns.[63] In Europe, too, tobacco was initially believed to cure cancer,[64] as well as offer many other restorative properties for an individual's constitution, including invigorating the spirits or helping expel excess phlegm.[65]* But tobacco could also be purely recreational. When Columbus and his crew arrived in the Americas, they witnessed the Taíno people

---

* Although nicotine itself is toxic to humans at high doses, the harmful effects of smoking mainly come from the other substances in the tobacco and their combustion products, including various carcinogens, carbon monoxide and tar. Curiously, there are a handful of diseases, including Parkinson's disease and fibroids, that appear to be negatively correlated with smoking – the habit might actually help prevent them. But this protective effect is far outweighed by the increased risk of developing smoking-related diseases.[66]

walking around with tobacco pouches strung around their necks, always on hand for a quick smoke, in the way someone today might leave home on a night out with a pack of twenty shoved in their back pocket.[67]

American indigenous cultures consumed tobacco in many different ways, often reflecting environmental conditions. Across Central and North America, tobacco was most commonly consumed by curing the leaves and smoking them as cigars or in pipes; the latter could be impractically large, highly decorated implements befitting their spiritual, and communal, usage. In the wetlands of the Amazon basin, where kindling a fire was trickier, tobacco was made into a drink; whereas in the thin air high in the Peruvian Andes, where smoking leaves you a little breathless, powdered tobacco snuff was snorted instead.[68] It was also chewed and held pressed against the gum as a wet wad of leaves or applied as eye drops. The Maya even administered it as an enema,[69] with the bulb made from an animal bladder and the shaft fashioned from the hollow femur of a small deer.[70] (Unsurprisingly, this last method never took off with Europeans.)

Europeans first encountered the plant when Columbus made landfall in Cuba, where the friendly islanders presented the visitors with food and exotic fruits they had never seen before. The gifts also included some dried tobacco leaves, but the puzzled Spanish found them utterly inedible and threw them overboard.[71] They realised that they weren't intended to be eaten, when they saw them being rolled into tubes, lit at one end, and smoked. Smoking was a totally novel experience for Europeans; while they burned incense in their churches, they did so for the scent and not to breathe in the smoke.

Early explorers in the New World even struggled to find the words to describe the practice. The Dominican friar Bartolomé de las Casas, one of the first European settlers in the Americas,[72] records how messengers sent ashore in Cuba found 'men with half-burned wood in their hands and certain herbs to take their smokes, which are some dry herbs put in a certain leaf, also dry, . . . and having lighted one part of it, by the other they suck, absorb, or receive that smoke inside with the breath, by which they become benumbed and almost drunk, and so it is said they do not feel fatigue. These, muskets as we will call them, they call *tabacos*.' Las Casas commented, 'I knew Spaniards on this island of Española who were accustomed to take it, and being reprimanded for it, by telling them it was a vice, they replied they were unable to cease using it.'[73]

Not only was this the explorers' first contact with smoking as a practice, therefore, but they also encountered the gnawing urge of chemical dependency: nicotine is far more addictive than alcohol, for instance. A century later, the philosopher and scientist Sir Francis Bacon similarly observed, 'In our time the use of tobacco is growing greatly and conquers men with a certain secret pleasure, so that those who have once become accustomed thereto can later hardly be restrained therefrom.'[74]

Consuming tobacco, in whatever form, delivers nicotine into the bloodstream, from where it is rapidly transported to the brain. Just as caffeine acts as a chemical mimic of the neurotransmitter adenosine, nicotine has a structure similar enough to another of the brain's signalling molecules, acetylcholine, to bind to acetylcholine receptors on the surface of neurones.[75] This triggers a cascade of other neurotransmitters into action, including dopamine in the mesolimbic reward pathway, producing the pleasurable effects of tobacco.[76] Inveterate tobacco-users keep their brain in a continual bath of nicotine, which alters its chemistry as the neurones adapt to the supply of the drug. The initial psychoactive rewards diminish as this tolerance builds, and nicotine use is continued mostly in order to avoid the negative sensations of withdrawal from the drug, such as irritability and anxiety. As with other addicts, when they try to quit, smokers (and more recently vapers) are held hostage by the unpleasantness of the withdrawal symptoms.

Rodrigo de Jerez, one of Columbus's messengers in Cuba and so one of the first Europeans to observe smoking, caught the habit himself and brought it back to Spain. His contemporaries were so horrified by the sight of smoke pouring out of his mouth and nose, making him look like a fuming demon, that he was locked up by the Inquisition for seven years.[77] But even so, by the 1530s, cigar smoking had become popular in Spain and Portugal. The French ambassador Jean Nicot, sent to Lisbon to arrange a royal marriage, picked up the habit and sent tobacco seeds to his queen. The 'Nicotian herb' quickly became wildly popular in the French court.[78] By 1570, botanists were referring to the weed as *nicotiana*, which also gave us the name for the active compound.

Genoese and Venetian merchants took tobacco to the Levant and Middle East, the Portuguese brought it to Africa, and before long, global maritime trade delivered tobacco to India, China and Japan.[79] So, while Columbus never realised his goal of reaching the Orient, the compelling weed he discovered on his voyage did make it to China.

Europeans adopted different means for consuming tobacco, depending on how they had first encountered its use. Cigar smoking became popular in Spain from the early sixteenth century, but later that century, Englishmen preferred to puff on a clay pipe, a habit they copied from North American natives in what became Virginia and the Carolinas. In the Middle East, the pipe was adapted into the waterpipe, or hookah, which cooled the smoke before inhalation and also emphasised smoking as a shared, communal activity.[80]

Tobacco therefore spread rapidly around the world once it had been discovered by European explorers. With its mind-altering chemistry it also became a key factor in the colonisation of the eastern seaboard of North America, with tragic and long-lasting consequences.

THE SUCCESS OF VIRGINIA

Tobacco was relatively slow to reach England, probably not arriving there before the second half of the sixteenth century. The privateer and slave trader Sir John Hawkins is believed to have made the first delivery of the leaf, captured during his raids of the indigenous peoples along the Florida coastline. Other swashbuckling pirates, such as Sir Francis Drake, targeted Spanish colonial settlements in the Americas and preyed on their shipping to loot tobacco and other treasures. But tobacco's most ardent advocate in England was Sir Walter Raleigh, who used his standing and influence at the court of Queen Elizabeth to promulgate its virtues. Very soon smoking was all the rage within the upper classes, and rapidly propagated throughout the population. And smoking was not yet another short-lived fad: nicotine's addictiveness meant that the habit grew and persisted, perniciously permeating the fabric of society.[81]

With no permanent presence in the Americas, and instead relying on pillage and piracy to enjoy the treasure they offered, the problem was that England had no reliable source of the plant. The little tobacco grown on home soil was deemed grossly inferior to that nurtured in the tropical sun of the West Indies.[82] Raleigh attempted to found the first permanent English settlement in North America on an island just off the coast of Virginia, its name honouring the presumed virtue of the Queen. The first Roanoke colony failed, and a second settlement was found mysteriously abandoned in 1590. While the Spanish had spent the past century consolidating an expansive empire across the New World, the English (and other European powers) were still struggling to establish a single successful colony.

Another attempt was made shortly after the new king, James I, made peace with Spain in 1604 after a generation of war. Yet the crown was wary about funding any more colonial ventures after the failure of Roanoke, and so the next expedition would be privately financed. The Virginia Company was chartered by the king in 1606, and after funds had been raised by selling shares and willing settlers recruited, Jamestown was founded the following spring on the banks of a major river spilling into the mouth of Chesapeake Bay.*

* The mouth of Chesapeake Bay was not only the location of the first successful English colony in North America; perhaps fittingly, it was also where

Yet the early years were tough. Over half of the settlers died within the first twelve months. Two relief convoys brought further colonists and supplies, but by 1610, 80 per cent of the settlers had perished, most succumbing to malaria and other diseases as well as hunger, especially during the previous winter, which became known as the 'starving time'.[83] A third supply fleet to Jamestown was scattered by a hurricane during the crossing, with the flagship badly damaged and only just able to make it to an uninhabited island in the Bermudas. Here, crew and passengers were marooned for ten months while they built two small ships to carry them on to their destination. (Their adventures were immortalised by Shakespeare in *The Tempest*, which is based on the real-life story of the *Sea Venture*.)[84] When they arrived at Jamestown, however, they found the settlement in ruins, with a surviving population of only 60 people. It was decided to abandon the colony and sail home with the remaining settlers. Yet in what proved to be a case of fortuitous timing, they had just set sail when they met another resupply fleet and so returned to the settlement. Now, with more supplies and a fresh influx of colonists, Jamestown seemed to be on a healthier footing.[85] But the Virginia colony still struggled to generate a profit for its backers: so far it had produced nothing suitable for export. No gold or silver had been found, and various ill-conceived farming attempts – including olive groves, vineyards and even the cultivation of silkworms – had failed.[86] The Virginia Company made it clear to the settlers that its financial backers expected a return on their investment and if the colonists didn't start delivering, they would be cut off. The future of the Jamestown settlement was now hanging by a thread.

But then, John Rolfe, who had been one of the castaways on Bermuda before making it to Jamestown, turned his hand to a potential cash crop. The colony had already been growing small

----

almost two centuries later the British were decisively defeated in 1781 and the United States earned its freedom from imperial subjugation. Jamestown and Yorktown are barely 20 kilometres apart as the crow flies.

amounts of an indigenous species of tobacco, *Nicotiana rustica*, but its smoke was harsh and there was little demand for the weed in England. The English much preferred the milder, sweeter leaf of *Nicotiana tabacum*, which they imported at great expense from the Spanish-controlled West Indies. Rolfe had been able to procure seeds of *N. tabacum* from Trinidad and spent several years trying to master the cultivation of the plant in the Virginia soil and climate, as well as the curing process of the harvested leaves.[87] In the meantime, he had married Pocahontas, the teenage daughter of the chief of the Powhatan tribe, as part of a deal to seal a fragile peace and trade agreement with the indigenous population.

When Rolfe's first tobacco crop was shipped back to London in 1613, it was well received – and, crucially, it turned a tidy profit. The English market was hungry for high-quality tobacco to feed its addiction. Jamestown quickly went from bust to boom. In 1618, the colony was told that it would no longer be financed by the Virginia Company but by its own tobacco farming. That year, it shipped 20,000 pounds of tobacco, rising to 60,000 pounds in 1622, 500,000 pounds in 1627, and 1.5 million pounds in 1629.[88] By the 1660s, Virginia was exporting an astonishing 25 million pounds of tobacco leaf every year.[89]

By the time Rolfe died in 1622, he had permanently reversed the fortunes of Jamestown. It served as the colonial capital until the end of the seventeenth century, by which point Virginia had a burgeoning population of nearly 60,000 settlers.[90] Tobacco continued to sustain and drive the growing English colonies in Virginia and Bermuda throughout the seventeenth and eighteenth centuries before sugar, coffee and cotton provided an additional economic base. As the colonisation of North America progressed, Virginia tobacco was not just exported back to England but sold throughout the eastern seaboard and the Caribbean. Indeed, the Thirteen Colonies came to use the sought-after leaf as a currency to trade with their neighbours – tobacco was not just a lucrative cash crop, it was literally used as cash.

Thus John Rolfe had planted the seeds for the creation of English America. Having come so perilously close to utter failure and abandonment, it was tobacco and its addictive properties that had secured the success of Jamestown – and ultimately the predominance of English language, culture, laws and other institutions in what would eventually become the most powerful nation on Earth.

The cultivation of tobacco in the Virginia colony had three profound and far-reaching consequences for history. First, the focus on tobacco as a cash crop marked a critical shift from subsistence farming to an agrarian economy in the colonies, making them commercially viable and self-supporting, while generating a handsome profit for the financial backers of these colonial ventures at home. And as tobacco had made the first English colony in the Americas viable, it paved the way for further expansion and settlement, with the colonies becoming enticing destinations, pulling more and more colonists towards North America.

Second, the tobacco plant is a hungry crop, drawing lots of nutrients out of the ground while it grows.[91] This means that tobacco cultivation quickly depletes the soil, and after just three years the ground must be left to recover for a decade or two[92] – or fresh soil be brought under the till. By the end of the seventeenth century, an estimated half a million acres of Virginia had been deforested and cleared, largely for tobacco plantations.[93] Tobacco was therefore a powerful driver for constant expansion and the settling of new lands to the west. This brought the colonists into direct conflict with the indigenous peoples living there, leading to mounting hostilities and ending in massacres and expulsion of the native tribes.

Third, the cultivation of tobacco was extremely labour-intensive. In the early years, the colonies relied on indentured labourers: settlers were offered passage to the New World but on arrival had to pay off this debt by working on the plantations for five to seven years. They were free when their service was finished, and as part of their contract were often promised

Changing our Minds   193

a grant of land. But neither this supply of indentured labourers nor the transport of convicts to the American colonies was able to meet the rapidly growing demand for plantation workers – particularly with the heavy burden of diseases in the region, as we saw in Chapter 4.[94] It was cheaper to import slaves from Africa, who were already hardened against tropical diseases, could be forced to work without a fixed-term contract and existed on a level of bare subsistence. Thus tobacco in North America, like sugar in the Caribbean, drove the expansion of the transatlantic slave trade in the early seventeenth century.

The *Nicotiana* herb remained central to the American economy, with the big tobacco companies emerging as some of the largest corporations in the country, and also able to wield great political power. This central position within the American economic and political development is plain to see in the Capitol building, the home of US Congress, where every Corinthian column in the grand Hall of Columns is topped with tobacco leaves, figuratively holding together the entire institution.[95]

Tobacco continued to be consumed by the ancient methods – chewed, snorted or smoked in pipes and cigars – but after 1880 a new delivery method for the nicotine fix became available: a pack of cigarettes. Previously, cigarettes had been crafted by hand from a pouch of tobacco and thin papers on an ad hoc basis – 'rollies' that you may be familiar with from your student days. Buying pre-rolled cigarettes was prohibitively expensive as a worker could hand-roll only a few a minute. The shift came with the invention in Virginia of a way to industrialise the process – a cigarette-rolling machine that could spit out over 200 cigarettes every minute. But initially, demand remained low for pre-rolled cigarettes, which were still perceived as pricey luxury items. At the dawn of the twentieth century, cigarettes only accounted for around 2 per cent of tobacco consumption in America. This all changed with unprecedented budgets spent on advertising, including campaigns after the First World War that were aimed at women.[96] Crucially, governments supplied their soldiers with cigarettes to keep up morale in the First and

Second World Wars, thereby creating legions of loyal (addicted) customers for the cigarette manufactures in peacetime.

THE UNREASONABLE EFFECTIVENESS OF ALKALOIDS*

As we have seen, plant products such as caffeine and nicotine became widespread around the world because they produce a pleasurable buzz and are addictive. They mimic signalling molecules in our brains and activate the mesolimbic reward pathway, an evolutionarily ancient system whose function is to motivate and tune the behaviour that would have aided our survival and reproduction in our natural environment. But why are such plant compounds able to hack the human brain's signalling system so effectively? Why would an otherwise unremarkable shrub growing natively in a pocket of the Ethiopian highlands, for example, just so happen to produce such a profound effect on our neurochemistry and so influence human history?

Caffeine, nicotine and morphine (which we'll come to) are all examples of alkaloids. These are organic compounds mostly made up of carbon atoms, some of which are arranged into rings, and containing at least one nitrogen atom.[97] They are an incredibly diverse family of natural compounds – around 20,000 are known – many of which are produced by plants to protect themselves from herbivorous animals,[98] which also explains why most of them taste bitter.

Alongside caffeine, nicotine and morphine, many other compounds we encounter in this book, such as quinine, cocaine,

---

* The title of this section is a hat-tip to an influential paper written in 1960 by the theoretical physicist and Nobel laureate Eugene Wigner, 'The Unreasonable Effectiveness of Mathematics in the Natural Sciences'.[99] In it, he discusses how uncannily effective the language of mathematics turns out to be at describing the physical reality of the universe, despite there being no discernible reason why this should be the case.

codeine and mescaline, are also alkaloids. (Their names are a giveaway: the naming convention for alkaloids is to slap the suffix '-ine' onto the formal scientific name of the plant it was extracted from.) In fact, alkaloids possess a staggering range of medical effects on the human body. They serve as anti-inflammatory and anticancer drugs, analgesics and local anaesthetics, and muscle relaxants. They suppress abnormal heart rhythms, widen or constrict blood vessels, lower blood pressure and reduce fever – alongside exerting stimulant or hallucinogenic effects on our brain. So what is behind this disproportionate potency of alkaloids in influencing the human body?

The answer is partly a numbers game. The botanical world produces a staggering number of different compounds, and so you might expect, purely by chance, some of them to incidentally have an effect on human biochemistry. But there's more to it than this. Unlike animals, plants can't move to run away from a threat or find resources they need. They're rooted to the spot and instead need to rely on their internal biochemical machinery to manufacture a diverse repertoire of chemicals to influence animal behaviour. These compounds serve to attract animals to the plant, perhaps to help with pollination, or to deter an insect herbivore, such as a caterpillar or beetle, chewing on its leaves. They are also deployed in the competition with other plants or against attacks from fungi.

Over evolutionary time, plants have become extremely adept at producing compounds that have a specific effect on animals. Many characteristics are shared between animals, because we descend from a common ancestor and certain features are conserved by evolution if they serve an important function. This applies not just to the physical design of organs but also to the precise structure of the molecules that allow our cells to communicate or run vital biochemical processes. So a plant molecule that has been designed by evolution to have a powerful effect on, say, an insect's digestive, circulatory or nervous system, is also likely to have a similar effect on the human body. Still, the alkaloid-producing plants that have had by far the greatest impact on human history are those that affect the mesolimbic pathway in our brain, such as caffeine, nicotine and, as we'll see shortly, morphine.*

When early humans noticed the intoxicating effects of these plants, they began seeking them out and cultivating them. Today, stimulant crops occupy an enormous area of farmland. Coffee is grown on roughly 10 million hectares worldwide, and tea and tobacco about 4 million hectares each. Together, this constitutes well over half the total land area used for growing all the rice in China.[102] And all this for crops that provide no nutrition for hungry mouths, nor fibres for clothing – just molecules with mild psychotropic effects.

In this sense, the nicotine synthesised by the tobacco plant to defend itself from insect herbivores[103] turned out to be a stunningly effective adaptation for propagating itself around the planet by exploiting human neurochemistry. Countless people worldwide are dedicated to nurturing the plants in the field, tending to their every need: keeping them carefully watered and nourished with concentrated fertiliser; weeding and removing other plants competing for light; and eradicating pests that may attack the plant. Seen this way, we might wonder who has really domesticated whom.

---

* Other plants also produce psychoactive alkaloids that are able to activate the mesolimbic pathway. For example, for several millennia, people across Peru and Ecuador have chewed coca leaves, usually along with a little mineral lime to help release the cocaine molecule into their bloodstream. The stimulant effects of coca leaves, either chewed or boiled to make a tea, alleviate hunger and fatigue. Incas cultivated the crop on a large scale. Coca Cola derives its name from the coca extracts used in the formula from 1885 until about 1903 (and the kola nuts used as a source of flavouring and caffeine). Cocaine works by blocking the molecular pump that removes dopamine once it has been released into the synapses between neurones, and so keeps the levels high.[100] It becomes much more potent, and thus addictive, when it is processed and concentrated into a powder that is snorted, heated and inhaled or dissolved and injected.[101]

OPIUM

As we saw earlier, the demand for tea in Britain grew steadily through the eighteenth century. Yet by the 1790s, most of it came from China, with the East India Company shipping 23 million pounds of tea leaves from the Far East to London every year.[104] The profits earned by the company for its shareholders were enormous, but the tea trade was also very lucrative for the British government – in the early 1800s, tea was taxed at 100 per cent of its value.

But there was one major problem: China didn't want anything the British Empire could offer in return. It had little interest in any of Britain's raw materials or industrially manufactured goods. China bought a little metal and some mechanical novelties,[105] but not on the scale needed to balance the volume of tea imported to Britain. The Qianlong Emperor wrote to King George III in 1793, 'Our Celestial Empire possesses all things in prolific abundance and lacks no product within its borders. There was therefore no need to import the manufactures of outside barbarians in exchange for our own produce.'[106] Britain was facing a colossal trade deficit.

The only European commodity that China desired was hard cash in the form of silver. Throughout the second half of the eighteenth century, therefore, about 90 per cent of Britain's trade exports to China were bullion.[107] The British government was struggling to raise enough silver to keep this trade going, and the East India Company was becoming concerned about maintaining its profits. At first, the company was able to operate a three-step triangular trade system: British, industrially manufactured goods were shipped to India, Indian cotton to China, and Chinese tea back to Britain, with the East India Company profiting handsomely on each leg. (This was a similar system to the transatlantic triangular trade that had operated from the sixteenth century, successively transporting European manufactured goods, African slaves and cash crops grown on

the American colonial plantations such as sugar – or rum distilled from it – tobacco and, later, cotton.) While the direct
exchange of tea for silver worked out badly for Britain, the
three-stage cycle neatly exploited disparities in supply and
demand in different commodities around the world. But the
Chinese demand for imported cotton waned, and once again
England was haemorrhaging precious silver to the East.[108]

But then the shrewd agents of the East India Company realised they could create a growing market for something that they
could source in bulk. While the Chinese government would only
consider silver for official trade, the Chinese people were keen
on something else: opium.

Opium is the latex fluid exuded from cuts made in the imma-
ture seed capsules of certain varieties of poppy, which is then
dried to a powder. This latex contains the analgesic compound
morphine (and also codeine) that provides pain relief and produces a warm feeling of relaxation and detachment. Poppies
were cultivated for their opium in Mesopotamia by the Sumerians from the end of the third millennium BC and named 'plants
of joy'. Opium use continued in the Middle East, as well as in
Egypt, and the drug was known in ancient Greek medicine at
least as early as the third century BC. By the eighth century AD,
Arab traders had taken opium to India and China, and between
the tenth and thirteenth centuries it made its way across
Europe.[109]

Taken orally, opium was used medically to treat pain. The
morphine is able to bind to nerve cell receptors (which are normally targets for the body's own hormones such as endorphins)
in parts of the brain involved in the sensation of pain, such as
the thalamus, brainstem and spinal cord. But opiates also bind
to receptors in the mesolimbic reward pathway, and so beyond
its medicinal properties, opium was therefore craved as a recreational drug.

Opium was legal in Britain in the early 1800s, with Brits consuming between ten and twenty tonnes of the stuff every year. [110]
Powdered opium was dissolved in alcohol as a tincture called

laudanum, which was freely available as a painkiller and even present in cough medicine for babies. Many late-eighteenth and nineteenth-century literary figures were influenced by opium, including Lord Byron, Charles Dickens, Elizabeth Barrett Browning, John Keats and Samuel Taylor Coleridge; Thomas De Quincey found fame with his autobiographical *Confessions of an English Opium-Eater*.[111] Drinking opium in this way produced mild narcotic effects but was also habit-forming – society at this time was therefore pervaded by high-functioning opium addicts, including many among the lower classes who were looking to numb the tedium of working and living in an industrialised urban world.[112] But while laudanum helped inspire a few poets and fuelled bouts of aristocratic debauchery, drinking it delivered a relatively slow release of opiates into the bloodstream.[113]

The Chinese, on the other hand, had taken to smoking opium. This delivers a much more rapid hit, which is consequently far more potent and addictive. The Chinese probably first came across opium smoking in the seventeenth century in the Dutch colonial outpost in Formosa (Taiwan); the Portuguese then began shipping the drug from their Indian trading hub in Goa to Canton in the eighteenth century.[114] So, although the East India Company didn't create the initial demand for opium in China, they certainly hammered a wedge into this crack to pour in the drug. They could bank on the property of addictive substances: once you've gained a clientele for your product, you can be assured that your customers will keep coming back. Instead of sending silver to China, the East India Company trafficked opium – and they could effectively grow as much of this new currency as they needed.[115] Before long, the company was pushing the drug in amounts never seen before. Ultimately, it boiled down to one addiction being traded for another – caffeine for opium – but the British were forcing a far more destructive substance on the Chinese. In order for the English mind to be focussed with tea, the Chinese mind was fogged with opium.[116]

The East India Company had seized control of Bengal from the Mughal Empire after the Battle of Plassey in 1757. It came to establish a monopoly on opium cultivation in the region and started running the drug into China. Opium consumption for non-medicinal uses was outlawed in China – the first laws banning opium had been enacted in 1729[117] – and so the East India Company couldn't be seen to be illegally importing opium as that would force a response from the emperor. Instead, it used independent 'country firms' as middlemen – Indian merchants licensed by the company to trade with China. These firms sold the opium for silver in the Pearl River estuary, where it was then smuggled ashore. This was a thinly veiled effort by the company to wash their hands of their formal involvement in the trafficking. As historian Michael Greenberg has put it, the East India Company 'perfected the technique of growing opium in India and disowning it in China'.[118] Meanwhile, a network of opium distribution spread through China, helped by corrupt officials who'd been paid off to look the other way.

The East India Company readily expanded its pipeline pumping opium into China until, in 1806, the tipping point was reached and the trade deficit had been forcibly reversed. The large numbers of Chinese opium addicts were now collectively paying so much to feed their habit that the import revenue from the smuggled opium exceeded that of the British tea exports. The silver tide had been turned and the precious metal began flowing from China to Britain for the first time.[119] The amount of opium imported into China by the East India Company trebled between 1810 and 1828, and then almost doubled again by 1832, to around 1,500 tonnes every year.[120] The British Empire, fuelled in the early days of its expansion across the Atlantic by one addictive plant, tobacco, was now wielding another, the poppy, as a tool of imperial subjugation.

We may never know for sure just how many Chinese men (it was mostly a male habit) were addicted to opium by the 1830s, but estimates at the time ranged between 4 and 12 million.[121] Although opium did destroy the lives of those heavily addicted – transforming them into stupefied zombies when high and at all other times listless and craving their next visit to the opium den – it didn't directly harm a large fraction of the whole population. The drug remained relatively expensive and so largely limited in accessibility to the mandarin and merchant classes in China.[122] The catastrophe for China wasn't so much the public health consequences as the economic disruption. As the silver paid to the British opium traffickers flowed out of China, the domestic supply diminished and the value of the precious metal rose. A farmer who had never touched an opium pipe now had to sell more of his crops to raise enough silver to be able to pay his taxes. By 1832, taxes were twice as high in real terms as they had been 50 years earlier[123] and the outflow of bullion was having a direct impact on the Chinese imperial treasury.[124] There was moral outrage in China against British opium and its harmful health effects, but the imperial court was far more concerned by the fiscal devastation it was wreaking.[125]

Some mandarins urged the emperor to legalise the drug and undercut the price of the British imports, or to prescribe that it only be traded for tea, so as to staunch the flow of silver out of the country.[126] But as a good Confucian, the Daoguang Emperor wanted to help save his people from themselves. In 1839, he declared a war on the drug and appointed a high-flying and moralistic bureaucrat, Lin Zexu, to stamp out the opium trade coursing through the coastal Canton province, where the drug was landed by merchants at the port of Guangzhou.

When he arrived at the foreign trading post, or 'factory', in Guangzhou, Commissioner Lin peremptorily ordered the British and other foreign merchants to immediately cease selling opium and hand over all the stock they had in the port's warehouses to be destroyed. The traders refused, and in response Lin had the doors of the factories nailed shut and their food supply cut off.[127]

The chief superintendent of trade for the British in China, Captain Charles Elliot, attempted to diffuse the standoff. He was able to persuade the traders in Guangzhou to turn over a staggering 1,700 tonnes of opium from the port storehouses by promising that the British government would reimburse them for their losses. Lin had the seized opium – worth an absolute fortune – disposed of by mixing it with water and lime in huge pits and then dumping the sludge into the Pearl River. The drugs bust was so large that it took three weeks to destroy it all.[128] Commissioner Lin thought he was doing his honourable duty to stamp out the illegal smuggling of opium that was eating away at his countrymen; but the events that day would lead to a clash of empires and a humiliating defeat for China.

The deal struck by Elliot in Guangzhou seemed to have satisfied everyone: Lin successfully seized the drugs cache and destroyed the contraband; the traders accepted the offer of getting paid full price anyway; and Elliot defused the flashpoint and kept the port open to British trade. Everyone, that is, apart from the prime minister, Lord Melbourne, who soon learned that the superintendent in Canton had jauntily promised this huge payout on his behalf. The government now had to find £2 million (equivalent to £164 million today) to compensate the drug dealers.[129] A local drugs bust had become an international incident, not just affecting merchants but challenging national pride. Lord Melbourne felt backed into a political corner, left with no other choice but to use military action to force China to reimburse Britain for the destroyed goods.

The response was to become a common theme of European imperialism: gunboat diplomacy. A taskforce of 4,000 British troops and 16 ships was dispatched to China, in a conflict known as the First Opium War (1839–1842).[130] Within the Royal Navy fleet was a new kind of ship, the *Nemesis*: a steam-powered warship made of iron, unmatched by anything the Chinese possessed. The *Nemesis* proved a devastating weapon of the newly emerging industrial age of warfare, protected by her thick iron armour and blasting Chinese junks out of the water with large, pivot-mounted cannons and rockets.[131] She was also able to steam through shallow water and penetrate far up rivers that deeper-draught wooden ships could not access. The British fleet blockaded the mouth of the Pearl River at Canton and captured a number of ports along the coast, including Shanghai and Nanking.[132] On land, Chinese armies were torn apart by British rifles and military training. China had invented gunpowder and the blast furnace, but now a European imperial power was arriving on its shores turning these innovations against it.

In July 1842, British ships and troops effectively closed off the Grand Canal, a crucial artery distributing grain throughout China. Beijing was threatened with famine, and the Daoguang Emperor was forced to sue for peace. The Treaty of Nanking was humiliating. China was forced to pay huge reparations for the confiscated opium and subsequent conflict, cede Hong Kong (the 'fragrant harbour') to the British as a colony and open five 'treaty ports', including Canton and Shanghai, to British merchants and other international trade. But the British were still not satisfied, leading to the Second Opium War (1856–1860) and the greater opening of China to foreign merchants as well as the full legalisation of the opium trade.

Opium imports rose to a peak in 1880 – after which they were superseded by the Chinese-grown opium supply – when almost 95,000 chests, or some 6,000 tonnes, of the drug were shipped from India.[133] Recreational opium use expanded across China, extending from the urban elites and middle class to rural workers.[134] By the time Japan invaded China in 1937, 10 percent of the population – some 40 million Chinese – were believed to be addicted to opium. It was not until after the communist takeover in 1949 and the arrival of the totalitarian regime of Chairman Mao that rampant opium addiction was finally stamped out in China.[135]

China endured an opioid crisis lasting 150 years, forced upon it by corporate greed and imperial coercion. Today, over a quarter of a million hectares of land are used for opium poppy cultivation, the vast majority of it grown illicitly in Afghanistan. Afghan poppies account for virtually all of the heroin available in Europe and Asia, while most heroin in the US is supplied from Mexico. In a recent survey, about ten million people in the US self-reported the non-medical use of opioids, although this figure is likely to be an underestimate (the survey data does not include the homeless or institutionalised populations, for instance). However, over 90 per cent of this opioid consumption is not heroin but legally produced, painkilling pharmaceuticals misused by those who have become addicted to such medication.[136]

This current opioid epidemic, echoing that in nineteenth century China, is not a uniquely American problem, but the US healthcare system created the conditions for the crisis to flourish. The US does not provide universal healthcare paid for by taxes, and instead, medical treatment must be covered by health insurance, which often favours the cheaper option of a painkiller pill over alternatives like physical therapy.

In the late 1990s, drug companies, including Purdue Pharma, looking to increase the prescription of opioid drugs, and thus their profits, were able to convince the regulators and medical community in the US that their oxycodone pills (sold under the brand name of OxyContin, among others) were not addictive. Patients were prescribed ever-higher opioid doses as they built up a tolerance, until many developed a dependence and became reliant on the drug to avoid nasty withdrawal symptoms. Millions of addicts continued to seek opioids on the black market, and between 1999 and 2020, more than half a million died from an opioid overdose.[137]

The US Department of Health and Human Services declared a nationwide public health emergency in 2017, and steps are being taken to control the opioid crisis,[138] but overdose deaths from synthetic opioids such as tramadol and fentanyl continue to rise.[139] Opium itself is today a minor recreational drug, but humans are just as susceptible to the pleasurable – and addictive – properties of opioid compounds.

Chapter 7

Coding Errors

This disease, so frequently attending all long voyages, and so particularly destructive to us, is surely the most singular and unaccountable of any that afflicts the human body.
—Richard Walter, *Anson's Voyage Round the World*

The complete instruction manual for building and operating a human is contained within the genome – the sequence of DNA in our cells. This genome consists of some 3 billion base pairs – the letters of the genetic code – arranged over 23 pairs of chromosomes, like information arranged in the separate volumes of an encyclopaedia. [1] This full DNA complement is held within every one of our cells, except for those with no nucleus, such as red blood cells, and sperms and eggs, which contain only one set of chromosomes. Every time a cell in our body divides, allowing us to grow or maintain our organs or heal a wound, the complete complement of genetic information is copied. This replication process is not 100 per cent perfect, however, and DNA can also become damaged (such as by radiation or certain chemicals). Errors, or mutations, are thereby introduced into the genetic code. The DNA alphabet consists of four letters – A, G, C and T (abbreviations of the bases adenine, guanine, cytosine, and thymine) – and mutations are often due to one of these genetic letters becoming replaced with a different one, like a spelling mistake.

One of the most infamous literary typos in recent years appeared in the historical fiction novel *The Queen's Governess* by Karen Harper. The protagonist has been abruptly woken after a night of passion: 'In the weak light of dawn, I tugged on the gown and sleeves I'd discarded like a wonton last night to fall into John's arms.' The intended word, presumably, was 'wanton', an archaic term for a promiscuous woman, rather than a kind of Chinese dumpling. The substitution of a single letter has completely changed the meaning of the word. In biology, a mutation in the genetic code can change one of the building blocks of a protein, reducing that protein's functionality or inactivating it completely.

Overall, the molecular machinery within our cells preserves the DNA code extremely faithfully. We possess hundreds of dedicated proteins for not only replicating DNA, but also proofreading the copies and repairing damage as it occurs. This means that the probability of any letter in the DNA code mutating is less than one in 10 million. But multiplied over the enormous number of base pairs we possess in our whole DNA complement, this error rate still equates to between 100 and 200 mutations arising in the genome of an individual that are passed on to the next generation.[2]

Most of these mutations have little to no effect on the individual. They strike somewhere within the huge expanses of the human genome that don't actually code for any proteins (99 per cent of our genome is non-coding). Or, if a mutation does arise in a gene, the substituted nucleotide may not affect the functioning of the protein that is manufactured using the instructions. But every now and then, a mutation has a deleterious effect. The affected protein stops working properly, and the individual suffers a congenital genetic condition. Albinism, for example, is caused by a mutation that blocks the production of melanin. Huntington's and Tay-Sachs diseases are both relatively common disorders also caused by mutations to a single gene.

We saw in Chapter 1 how genetics altered the history of the Habsburg dynasty; we'll now discover how a spontaneous genetic mutation that arose in an English queen came to impact some of the most powerful royal families in Europe, and thus the fate of the continent.

THE CURSE OF THE COBURGS

Queen Victoria had nine children, all of whom survived to adulthood, and forty-odd grandchildren. The queen was convinced that royal intermarriages were the best means to ensure lasting peace in Europe, and her strategic matchmaking meant that by the end of the nineteenth century, her grandson, the future George V, was related by blood or marriage to practically every other royal family in Europe.[3] Victoria's children and grandchildren became the kings, queens, emperors and empresses of Germany, Prussia, Spain, Greece, Romania, Norway and Russia, as well as important dukes and duchesses elsewhere. Victoria has therefore been called the 'grandmother of Europe'.

Her youngest son, Prince Leopold, born in 1853, was a delicate, fragile child. He suffered joint pain, and the slightest bumps and scratches would cause him huge bruising and a great deal of bleeding. He was permanently fretted over by royal physicians but survived to marry and have two children of his own. But at the age of thirty, he slipped down a staircase,[4] bumped his head and died of a cerebral haemorrhage.[5] Leopold's poor health may have been taken to be an unfortunate, isolated illness, but disturbingly, a similar trend was emerging among Victoria's other male descendants. People began whispering about the 'curse of the Coburgs'. Leopold, often unkindly referred to as the 'Bleeder Prince', was a harbinger of a calamity that was to strike royal families across Europe.

Today, the curse of the Coburgs is known as haemophilia.

Haemophilia is caused by a genetic mutation that reduces clotting factors in the blood. These factors help to plug open wounds with a tangled mat of fibrous proteins – a blood clot. If a haemophiliac suffers a cut or a broken blood vessel, the blood will flow for much longer before a clot becomes established – sometimes for days. Even a modest gash can lead to significant blood loss, and a gentle bump can rupture subcutaneous vessels that continue seeping into the surrounding tissue to create a large, painful bruise or even a bulging haematoma. Such internal bleeding is especially dangerous as a surgeon can only attempt to intervene by creating another open wound.[6] Haemophiliacs also have an increased risk of bleeding into joints and the brain.[78]

These clotting factor genes are stored in the DNA on the X-chromosome. Females receive two X-chromosomes from their parents, so even if they inherit one copy of a defective clotting factor gene, they should have a second, normal copy that can compensate and trigger healthy clotting. They are carriers of the mutation but don't generally suffer from the disorder. Males, however, have XY sex chromosomes and so an inherited mutation on the X chromosome is not masked by another working copy. A female haemophiliac would have to inherit a double copy of the faulty gene: her mother would need to be a carrier and her father also a haemophiliac – a statistical rarity compounded by the unlikelihood of a haemophiliac surviving to reproductive age. Thus it is almost exclusively males that are born with haemophilia, which they inherit via the X chromosome from their mother – genetically speaking, it is known as a sex-linked recessive disease.

Queen Victoria was distressed by the appearance of this disease among her descendants but protested that it did not originate from her side of the family. Indeed, none of her siblings or ancestors suffered from this strange, debilitating malady, but then neither too did the family of her husband, Prince Albert. As haemophilia was previously unknown in Victoria's family tree, it appears that a spontaneous mutation of the blood clotting gene occurred in either her father's sperm or her mother's egg before her conception.[9]

Two of Victoria's five daughters, Princesses Alice and

Beatrice, also inherited the faulty gene and became carriers of the disease. And then their daughters married into the royal houses of Spain and Russia and bore sons – heirs to two important thrones – with haemophilia.[10] In total, over twenty members of European royal families inherited the condition from Victoria.[11] The marriage arrangements that the queen had taken such pride in had come to spread a debilitating, and often lethal, genetic disorder across Europe's royal dynasties. Ironically, however, although the mutation responsible arose within Queen Victoria, the British monarchy was spared as the line of succession passed through Edward VII, who had not inherited the faulty gene – winning a 50/50 genetic coin toss in receiving the unaffected X-chromosome from his mother.

The most catastrophic effects of Victoria's deadly mutation were felt by the royal houses of Spain and Russia.

King Alfonso XIII of Spain unwittingly made one of the most catastrophic decisions of his reign when he married Victoria Eugenie of Battenberg, Princess Beatrice's daughter. Despite her good health, Victoria Eugenie was a carrier of haemophilia. The Spanish Embassy in London warned Alfonso of the risk that she harboured the debilitating 'royal disease' – indeed, two of the young princess's three brothers were afflicted with haemophilia – but at the time it was impossible to check, and the royal pedigree of the princess offered a great deal of prestige. Alfonso pushed aside any concerns and married Victoria Eugenie in the spring of 1906.[12]

A son was born to them the following year, named Alfonso after his father, but on the occasion of his circumcision it was discovered that the crown prince did indeed have haemophilia. Their second child, Jaime, was spared the blood disorder but at a young age developed mastoiditis – an infection within the skull – and became deaf and mute. A healthy son, Juan, was finally born in 1913. In all, Alfonso XIII and Victoria Eugenie had seven children together: two haemophiliac sons, one deaf and mute son, two daughters who were possible carriers, one stillborn baby and just one healthy son. The heir apparent,

Prince Alfonso, was bedridden for long periods at a time with severe haematomas triggered by minor injuries, and he largely remained out of the public eye. Many considered that the royal blood of Spain had been tainted by the British princess.[13]

Political crisis in Spain led to a coup d'etat in 1923 and then a dictatorship – consented to by the king – under General Primo de Rivera, who ruled the country until his resignation was forced in 1930. At this point, the king had become hugely unpopular and believed that he could only save the monarchy by abdicating in favour of his eldest son and heir, Alfonso. But public opinion rode against succession of the haemophiliac prince or his deaf and mute brother, Jaime, both of whom were regarded as unfit for the burden of the crown. Had the king had the courage, he might have proclaimed his healthy son, Juan, as his successor and so changed the course of events that were to unfold.

In the municipal elections of April 1931, monarchist political parties won only a tiny overall majority and suffered decisive defeats in many of the largest cities, which were seen as votes against the monarchy. King Alfonso and his family went into voluntary exile as the Second Spanish Republic was declared, and within two years the two eldest princes had both renounced their dynastic claims to the defunct throne.[14] While republican sentiment in Spain was strong, and the monarchy increasingly unpopular, the 'curse of the Coburgs' had further undermined the monarchy. (Although, this was not the end of the Spanish kingdom: General Franco reinstated the monarchy in 1947 and nominated his successor in 1975 as Juan Carlos I, Alfonso's grandson.)

RUSSIA

Queen Victoria's second daughter, Alice, married Louis IV, Grand Duke of Hesse-Darmstadt. They produced two sons, one of whom died of haemophilia as a toddler.[15] Their youngest surviving daughter, Princess Alix, met Nicholas Alexandrovich

Romanov, the heir to the Russian Empire, in Coburg at the wedding of her brother and their mutual cousin. Nicholas proposed, and they were wed in 1894, three weeks after the death of his father and his ascension to the throne. Alix was received into the Russian Orthodox Church and took the name Alexandra Feodorovna. The risk of her acting as a carrier of the royal disease was known, but by this point, the affliction was so prevalent among European dynasties that it was accepted as something of an occupational hazard.[16]*

The primary duty of a tsarina was to produce a male heir to the throne, but Alexandra's first four children were all girls. In August 1904, she finally bore a long-awaited prince, Alexei. The painful truth soon became apparent, however, that the little *tsesarevich* suffered from haemophilia, transmitted down the female line from his great-grandmother Queen Victoria. Alexandra coddled the young boy, lest he should fall and suffer internal bleeding that could so easily prove fatal. A sailor was detailed to accompany Alexei wherever he went to catch him if he stumbled and to carry him when he was unable to walk, as occurred frequently. Alexandra consulted numerous physicians, but treatment for haemophilia was far beyond the medical science of the time, and she became convinced that only a miracle could save her only son and heir to the Russian Empire. She became increasingly withdrawn and turned to prayer for his salvation.

It was in these desperate circumstances, in the summer of 1907, that a fraught Alexandra was introduced to a healer, the enigmatic Grigori Yefimovich. He is better known to history as Rasputin. The name probably derived from *rasputnyi*, meaning 'dissolute',[18] and he certainly lived up to it, developing a reputation for debauchery, drunkenness, coarse language and licentious

---

* It wasn't until 1913 that the first royal marriage was declined on the basis of the risk of haemophilia, when the Queen of Romania ruled out a union between her son, Crown Prince Ferdinand, and Olga, the eldest daughter of Tsar Nicholas II and Alexandra.[17]

indulgence. He dressed in a peasant blouse and baggy trousers, with greasy black hair that reached his shoulders and a long, unkempt beard – in short, he had the appearance of a dishevelled vagrant. But to contemporaries, Rasputin's most arresting feature was the penetrating intensity of his ice-blue eyes.[19] He presented himself as a mystic and a holy man* with an aura of spiritual authority, and he was supposedly gifted with extraordinary powers of clairvoyance and healing. While we might call him a charlatan, he seemed to genuinely believe in his own abilities.

What's more, Rasputin seemed to have a preternatural ability to soothe and calm the frequently distressed and sometimes hysterical Alexei. Alexandra persuaded herself that he was also able to ease the prince's physical pain and stop his internal bleeding. Perhaps there was some truth in this beyond the wishful thinking of a desperate mother. Rasputin was known to use hypnotism, and by stupefying his patient, the reduction in stress and lowering of blood pressure and heart rate could conceivably have had a beneficial influence.[21] Whether Rasputin actually had any genuine effect on the tsesarevich's medical condition was a moot point: the empress believed he could help, and that was all that mattered. As Alexandra's dependence on him grew, Rasputin became an ever more frequent presence at the Romanov court.

In October 1912, the ten-year-old tsesarevich suffered a particularly bad bout of internal bleeding after accompanying his mother on a jolting carriage ride. The court physicians were powerless to treat a large haematoma swelling in his groin, and the prince was given the last sacraments. The desperate Alexandra sent a telegram to Rasputin, who was at his home in western Siberia. He wired back with the prophetic message, 'The Little

---

* Rasputin's strange hybrid of mysticism and eroticism was influenced by the Khlysty sect, which believed that one should sin as much as possible so that repentance and salvation would be all the more resounding.[20]

One will not die. Do not allow the doctors to bother him too much.'[22] Within hours, the prince's condition began to improve.

Perhaps the arrival of Rasputin's telegraph had simply been a case of good timing. Alexei had already been haemorrhaging for several days, and his body may simply have been able to begin healing itself by that point. Perhaps Rasputin's insistence on keeping the doctors away actually did have a positive effect: the gaggle of physicians fretting around his bed may have been doing more harm than good, constantly prodding the swollen haematoma to check its progression.[23] Furthermore, they were likely to have treated him with aspirin for pain relief, and unbeknown to them at the time, the medicine's side-effect of thinning the blood would only have exacerbated the bleeding.[24] Whatever the reason, soon after receipt of the prophetic telegraph, the tsesarevich began making a miraculous recovery. Alexei's health was the future of the Romanov dynasty, and Rasputin had become indispensable in the eyes of the tsarina.

With the eruption of the First World War in July 1914, Tsar Nicholas II found himself dragged into a continent-engulfing war against his cousin-in-law, Kaiser Wilhelm II of Germany, who was himself fighting his own cousin, King George V. Queen Victoria's grand plan of ensuring peace in Europe through an intertwined web of related royals had failed.

Over the first year of the Great War, Nicholas lost some four million of his subjects.[25] In September 1915, after the disastrous early defeats, Nicholas travelled to the front to personally take supreme command of the Russian forces. He knew little about military matters, however, and the army continued to flounder. In his absence, Alexandra assumed control of domestic affairs from the imperial palace in Petrograd (St Petersburg). She devoted most of her energies to influencing politics and making recommendations for government appointments. It was not only those within the imperial court but also ministers and commanders of the armed forces who rose or fell depending on her favour. And whispering in her ear was the ever-present Rasputin, using her as a vehicle to further his own power.[26] Alexandra urged her husband to follow the advice of 'Our Friend', as she referred to Rasputin, on state matters, and Nicholas often acquiesced.[27] In the seventeen months of the tsarina's rule, until February 1917, Russia had four prime ministers, five Ministers of the Interior, three Foreign Ministers, three War Ministers, three Ministers of Transport and four Ministers of Agriculture. Competent officials were removed from post and replaced by obedient supporters, and the government's ability to function was undermined as officials rarely stayed in office long enough to master the demands of their post, all of which contributed to political instability and domestic turmoil.[28]

As Rasputin's influence within the Romanov court grew, so too did the rumours of his misdemeanours. Lurid stories about his drunken orgies and sexual exploits circulated widely. Worst of all was the widespread rumour, which Rasputin only encouraged, that the tsarina herself was among his lovers.[29] Alexandra's close connection with Rasputin became increasingly damaging and undermined the reputation of the royal court among the Russian people. Support for the Romanov dynasty was being eaten away from the inside. Nicholas's supporters urged him to cut all ties with this toxic man; but he wouldn't hear it. Nicholas was aware of the whisperings, but he would not remove Rasputin so long as his wife believed that he, and only he, could keep their son alive.[30] 'Better one Rasputin than ten fits of hysterics every day,' he is reported to have exclaimed in an unguarded moment.[31]

Alexei's condition was kept a secret from the Russian people. Whenever he was suffering a bout of bleeding and missed a public appearance, the Romanovs issued an excuse, but no one believed them, and increasingly outlandish rumours circulated about his absences. Honesty about the heir's haemophilia would have prompted questions over his ability to rule and the dynastic succession, but maintaining the wall of secrecy was even more damaging. Alexandra's shy and introverted disposition came across as cold and arrogant, and the official reticence on Alexei's affliction only made the situation worse, undermining the nation's respect for the empress and thus too the tsar and the crown.[32]

By the autumn of 1916, the tsar and his wife had become intensely unpopular among all classes of Russian society.[33] The public were scandalised by Rasputin's behaviour and apparent leverage over the tsarina. This was poisoning the monarchy's relations not only with the Russian people, but also with the traditional pillars of the state: the aristocracy, the government, the Church and the army.[34] By assuming command himself, the tsar was increasingly seen as personally responsible for the military defeats, and the flawed decision to take over was blamed on the influence of Alexandra and Rasputin. Some even muttered that Nicholas and his German wife were conspiring with the enemy.[35] Continued, grave losses at the front combined with food and fuel shortages at home, exacerbated by incompetent officials, turned the Russian people against the imperial family. The press, released from censorship after the 1905 Revolution, began to write openly of Rasputin as a sinister force within the palace who had the ear of the tsarina and used his influence to pull the strings like a puppet master. In the Russian parliament – the Duma – too, leftist politicians hinted at dark forces near the throne in their speeches.[36] As revolutionary sentiments bubbled, it mattered not what was true but what the people thought was true.

Tensions rose until, in December 1916, a plot was hatched by two members of the royal family and an ardent monarchist politician to end the insidious, corrupting influence of Rasputin once and for all. The mystic was lured to Yusupov Palace in Petrograd, where the conspirators poisoned him with cyanide, shot him twice in the chest and twice in the head, beat him with a candlestick, then finally, for good measure, pushed him beneath the ice covering the Moika River, where he drowned.[37] They hoped that with the 'Mad Monk' gone, the tsar could be persuaded to listen to sense, to relinquish command of the Russian forces to an experienced general, rule the nation in cooperation with the Duma and so save the monarchy.[38] But it was too late; the grave damage had already been done.

Within two months of Rasputin's murder, the Russian people's exhaustion from the war and long-running discontent with the tsar's rule erupted into mass protests and violent clashes with the police in Petrograd. It seemed that revolution now was a patriotic act to save Russia. For the rising Bolsheviks, Rasputin had been symptomatic of the broader corruption of the autocratic regime, and his murder by the nobility was seen as an attempt to hold on to power over the proletariat.

Both the Duma and the revolutionary Soviets tried to convince Nicholas to abdicate, allow Alexei to succeed as a constitutional monarch under the regency of his uncle Grand Duke Michael, and take the rest of the royal family into exile. But the tsar would not be parted from his haemophiliac son and declared he would only abdicate in favour of Michael. Whereas the people may have been satisfied with a twelve-year-old boy succeeding as tsar under the tutelage of the elected Duma, they would not accept a simple replacement with another imperial autocrat.[39] Michael delegated rule to the provisional government, and Nicholas and his family were placed under house arrest. A republic was declared in September 1917, but ongoing turmoil led to the October Revolution, during which the radical Bolsheviks seized power. The Romanovs were later moved to a house in the Urals, and in the early hours of 17 July 1918, Nicholas and Alexandra and their children, as well as the tsar's doctor and three servants, were executed by a Bolshevik firing squad. The Romanov dynasty had been extinguished.

It was a tragic ending to the Russian imperial family. We have seen how in Spain Prince Alfonso's haemophilia called into question his ability to succeed his father, weakening the monarchy in the years leading up to the establishment of a republic. The tsesarevich's affliction arguably had even more devastating consequences. The distraught tsarina was so desperate to believe that Rasputin could heal her son and heir to the empire, and Nicholas II too reluctant to remove him from the royal court, that the standing of the imperial family was damaged beyond repair. Rasputin's influence on state affairs, combined with unfounded rumours of an affair with Alexandra and collusion with the German enemy, eroded support for the government in Church and army, and fuelled public discontent until it eventually broke out in anti-monarchist rebellion. The gruelling war and food and fuel shortages stoked civil unrest, which made the Bolsheviks promising peace and bread so popular, but the consequences of royal haemophilia played a significant role too. After the October Revolution, the leader of the provisional government, Alexander Kerensky, concluded, 'Without Rasputin there would have been no Lenin.'[40]

And we could add that without a single chance genetic mutation originating with Queen Victoria almost exactly a century earlier, there would have been no Rasputin.

SCURVY

While a mutation inherited from Queen Victoria affected many of the royal families of Europe throughout the nineteenth and twentieth centuries, another genetic defect is present in all of humanity. A gene inactivated early in our primate evolutionary history manifests itself as the debilitating, and eventually fatal, condition known as scurvy.

Scurvy has been known throughout history, striking peasants during famines or armies during sieges. The ancient Egyptians recorded symptoms consistent with scurvy,[41] as did the Greek physician Hippocrates in the fifth century BC. The Crusaders suffered from scurvy, particularly when they were observing the restricted diet of Lent.[42] But the period in history when the affliction really reared its bleeding-gummed head was during the Age of Sail. From the end of the fifteenth century, advances in ship construction and navigation enabled mariners to undertake longer voyages across open ocean. This meant that hundreds of men were crammed into ships and spent many weeks, often months, away from land, with limited provisions.

The first recorded outbreak of scurvy at sea occurred during

Vasco de Gama's discovery of the passage to India via the southern tip of Africa. Struggling against the monsoon winds on their return trip across the Indian Ocean in 1498, de Gama and his crew spent three months continuously at sea. The disease unfolded in a characteristic, horrifying way. The sailors started feeling weakened and lethargic, with aching joints and bruising easily. Their hands and feet began to swell, then their legs, arms and necks, with their skin taking on a patchy, purple discolouration. Their gums became grotesquely swollen and bled, their teeth loosened and fell out, and their breath stank of putrescence. Fresh wounds failed to heal and became infected, and sailors who had sustained injuries years before watched in horror as the old wounds opened up again. The healed joins of previously broken bones also dissolved, the fracture spontaneously reappearing as if it had never fused. In the final stages of the disease, sailors suffered hallucinations or blindness, before eventually succumbing to it under great pain, their end often delivered by a catastrophic haemorrhage around the heart or brain. Of 170 crew members on de Gama's voyage, 116 perished, mainly of scurvy, and those who survived were in a dire state of health.[43]

For three centuries after this pivotal voyage, scurvy remained the greatly feared spectre that relentlessly stalked sailors at sea. Those that didn't succumb to it were lucky. Christopher Columbus and his crew were largely spared the ravages of scurvy because their journey to the Americas in 1492 took only 36 days, with the trade winds behind them the whole way. Magellan's first circumnavigation of the globe in 1519–1522 was less fortuitous: he lost 208 out of 230 sailors, again mostly to scurvy.[44] And this dreaded disease befell not just the history-making voyages we know today but every other long marine passage that didn't carry reliable countermeasures. It was the scourge of merchant ships plying the oceanic trade routes and of naval vessels duelling for control of the world's seas. Scurvy struck terror into the hearts of sailors through the chilling inevitability with which it appeared after only a few weeks away from land.

Throughout the great majority of the Age of Sail, a captain on a long voyage could expect to lose from a third to more than half of his crew to slow and agonising death by scurvy. Sir Richard Hawkins, an English explorer and privateer, described the malady in the early seventeenth century as 'the plague of the sea, and the spoyle of mariners'.[45] Over the three centuries of the Age of Exploration, between 1500 and 1800, it has been estimated that scurvy killed over two million sailors – accounting for more deaths at sea than storms and shipwrecks, naval battles and all other diseases put together.[46]

During the eighteenth century, at the height of the European power struggle for dominance at sea, the length of time that a navy could keep its sailors healthy and its warships combat-effective was a crucial factor in protecting trade routes, defending overseas colonies and ports and safeguarding against enemy invasion. The first sea power to learn how to effectively combat the debilitating effects of the disease would gain a decisive naval advantage, even if only in the short term.

We now know that scurvy wasn't caused by the often cramped and squalid conditions below deck on a sailing ship, nor by some feature of the open ocean itself, as had been supposed. Scurvy is a deficiency disease, caused by the absence of a specific ingredient in the sailors' diet. Their rations were restricted to foodstuffs that preserved well, largely salted meat or fish and cereal grains that were consumed as ship's biscuits or 'hardtack'.[47] Fresh fruit and vegetables were a luxury available only when making landfall. So, while sailors received a sufficient calorie intake to sustain their energy at sea, their diet lacked some crucial component required for the healthy functioning of the body.

A healthy human diet contains a balance of three key macro-nutrients: carbohydrates, fat and protein. We break these down during digestion and metabolism to supply the chemical energy that powers our body's activities and provide all the molecular building blocks from which we construct our cells. But we also need small amounts of other vital substances known as micronutrients. Minerals are inorganic compounds – mostly containing metals – that enable key processes in the body. Salt, for example, offers both the sodium needed by our nerves and muscles and the chlorine that is used for maintaining the water balance inside our cells and for making (hydrochloric) stomach acid. We also need calcium for building bones and teeth, iron for carrying oxygen around in our bloodstream, phosphorus and sulphur for synthesising other crucial components of our cells and smaller amounts of other metals such as copper, cobalt, iodine and zinc. These vital elements within minerals cannot be biochemically synthesised by living organisms – they ultimately derive from the soil and water taken up by the plants and animals we eat.

Other essential micronutrients are organic compounds known as vitamins, which, unlike minerals, have been synthesised by organisms. Humans need these specific chemicals to function properly, but we are biochemically incapable of manufacturing them ourselves and so must derive them from our diet. All chemical reactions inside cells are driven by specific enzymes, and if a species evolves with a mutation that has knocked out a metabolic enzyme, it may lose the ability to synthesise a particular chemical product (as well as other compounds derived from it). Different species would therefore define their own list of vitamins differently: while one may need to source a particular organic compound as an essential dietary micronutrient, another might be able to happily create their own. For humans, there are thirteen essential vitamins. These were named alphabetically as they were discovered over the twentieth century, with deletions from the list when it was found that a chemical wasn't actually required in the diet, and numbers added when scientists realised that some were chemically related to each other: they are vitamins A, B1/2/3/5/6/7/9/12, C, D, E and K. The human biochemical factory is able to synthesise vitamin D using a chemical reaction driven by UV rays in the sunlight hitting the skin. But many people living at higher latitudes cannot synthesise enough this way, so it is still considered a dietary vitamin.

The different vitamins are involved in a range of processes in the body, from assisting the action of enzymes in running vital biochemical reactions in our cells to helping us extract energy or absorb other key nutrients from our diet. If our diet lacks one of these crucial ingredients, our body's stockpile of it becomes depleted and we develop a deficiency disease.

Humans appear to be more flawed in this way than other animals.[48] Whereas most animals can spend their whole life eating just a single type of food without suffering any ill effects – buffalo are very happy munching nothing but grass, for instance – humans must secure an adequate supply of the numerous essential micronutrients from a particularly diverse diet.

Indeed, the very reason we have accumulated these metabolic mutations is because over our evolutionary history we have eaten a wide variety of plants – and more recently also scavenged and then hunted for meat – across a range of different habitats. Any mutation that knocked out a metabolic enzyme may not have had any immediate detrimental effect: the organic molecule it had produced was likely still present in the varied foods we ate, and so the mutation was not removed from the population by natural selection. We accumulated defects in our biochemical factory *because* they were masked by the varied diet of our primate and then hunter-gatherer ancestors. This means that our ancestors being ecologically gifted with a rich and varied diet created a lasting requirement for such a diet in order to survive.[49] When humans developed agriculture, and our diet became focussed on a restricted number of staple cereal crops and cultivated fruit and veg, deficiency diseases started to appear.[50]*

* We also need twenty different essential amino acids to build the various proteins in our body, but are only capable of creating twelve of them ourselves.[51] Here it is interesting to note that traditional agricultural diets around the world all combine a staple cereal crop to provide calories with a type of protein-rich legume.[52] In South and East Asia, rice was paired with lentils or

For sailors during the Age of Sail, the vital element lacking in their diet was vitamin C. It is a water-soluble vitamin, which means that the body can store only a limited amount of it (compared to fat-soluble vitamins such as A and D).[53] This organic compound is present in sufficient quantities in a balanced diet that includes fresh fruit and veg, but not in the preserved provisions aboard a ship. From the moment they left harbour, the sailors' limited body stores of vitamin C began to deplete, until after a month or so at sea they began experiencing the effects of its deficiency.[54]

All animals that suffer from scurvy lack a particular enzyme called L-gulonolactone oxidase, or GULO for short, which performs the final step in the chemical manufacturing pathway for making vitamin C.[55] Between 60 and 40 million years ago, our branch of the primate evolutionary tree acquired a mutation in the GULO gene that stopped this important enzyme functioning.[56] But this mutation went unnoticed by evolution as our forest-dwelling ancestors had plenty of vegetation and fruit in their natural diet rich in vitamin C. Over time, the GULO code accumulated more and more errors so that today it exists in human DNA as no more than a ghost gene, like an irredeemably rusty component in a car engine.

The main role of vitamin C in the body is to support the synthesis of a long-stranded protein called collagen. Collagen is the most abundant protein in the human body and provides structural strength within connective tissue. Connective tissue serves as the foundation of the skin and gives it its elasticity. It holds our internal organs in place, and it lines blood vessels and nerves. Collagen also makes up tendons and ligaments, gives structure to muscles, serves as the scaffolding of cartilage and bone and is integral to the process of wound healing. In short,

---

soya beans; in the Middle East, wheat was complemented with chickpeas or fava beans; indigenous Americans ate maize with black beans or pinto beans; and a common combination in Africa was millet and cowpea.

if the body stops being able to form and maintain collagen, its very fabric begins to fall apart.

Nearly every other animal species on the planet is able to synthesise its own vitamin C, usually in the liver,[57] but humans – as well as other apes, monkeys and tarsiers – have lost this biochemical ability.* It was a chance mutation millions of years ago within a tree-dwelling ancestor that was to have such dire consequences for mariners during the Age of Sail.

THE SEARCH FOR A CURE

One of the worst health disasters at sea was the catastrophe that befell a Royal Navy campaign in the Spanish Pacific, led by Commodore George Anson. In 1739, England and Spain once again descended into open warfare. Before being sub-sumed into the wider European power struggle of the War of the Austrian Succession, the Anglo-Spanish conflict began over the two age-old rivals' efforts to dominate trade around Span-ish America.† After British successes around the Caribbean early in the war, Commodore Anson was charged with leading a squadron of warships to harry Spanish possessions on the Pacific coast. His mission was to attack vulnerable towns and to capture one of the enormously valuable treasure galleons as

---

* The other major branch of primates, which includes lemurs and lorises, are able to synthesise their own vitamin C and so cannot suffer from scurvy; but several other unrelated species also need dietary vitamin C,[58] including guinea pigs and fruit bats.[59] A Norwegian study into scurvy in 1907 picked the guinea pig for its test animal, which was fortuitous as it just so happens to be one of the very few species alongside humans that can actually suffer from the affliction.[60] When it comes to studying scurvy, it turns out that the guinea pig makes the perfect guinea pig.

† The *casus belli* pushed in parliament to justify declaring this war meant it was subsequently referred to with perhaps one of the most ridiculous names for any conflict—'The War of Jenkins' Ear'—after a British merchant captain was mutilated by a Spanish coastguard who had boarded his ship off the coast of Florida and accused him of smuggling.

it transported silver from the mines in Mexico to the Philippines and then on to China for trade. After a lengthy period of preparation needed for repairing the ships and trying to muster enough men, Anson's squadron finally set sail in September 1740. But these months of delay, during which the assembled crew had already been living on ship's rations, meant that Anson's men were beginning to weaken with malnutrition even before they departed.

After they rounded Cape Horn at the tip of South America and entered the Pacific, Anson's ships were damaged and scattered during three months of fierce storms; two returned home and another was wrecked off the coast of Chile. But far worse than the raging of the winds were the ravages of scurvy. By the end of April 1741, practically every man aboard Anson's flagship, the *Centurion*, was suffering. Forty-three crew members died of the affliction that month, and double that number in May. As the number of men fit and able to work the ship's rigging rapidly dwindled, the scene below decks became increasingly desperate. The stench was unbearable amid the tightly packed hammocks full of the sick and dying. Corpses were frequently left where they lay, the surviving crewmen too weak to tip them overboard.

The *Centurion* was able to rejoin the two other remaining ships from the expedition at a pre-arranged rendezvous at the Juan Fernández islands.* Of the roughly 1,200 men who had set sail aboard the three ships ten months before, almost three-quarters had died. Despite now recuperating on dry land, with plenty of fresh crops planted there during previous expeditions, it took three weeks for the continuing deaths from scurvy to stop. After three months of recovery, the remaining ships continued their mission, attacking Spanish shipping and towns along the South American coast. In the summer of 1742, they

---

* It was here that a sailor named Alexander Selkirk had been marooned for more than four years at the start of the century, who is believed to have served as the inspiration for Daniel Defoe's novel *Robinson Crusoe*.[61]

Commodore George Anson's HMS *Centurion* captured a Manila galleon and circumnavigated the globe, but scurvy ravaged his crew on the long voyage.

set sail across the Pacific to China, again losing sailors every day to scurvy, until there was only enough crew for just one ship. Anson now pursued his second objective, sailing the *Centurion* towards the Philippines in the hunt for a Manila galleon, where in June 1742 they captured such a Spanish treasure ship, laden with over 1.3 million silver coins.

Anson finally returned home via the Indian Ocean, having circumnavigated the globe. After a voyage lasting nearly four years, the expedition had delivered a haul of loot, but its success had come at a staggering human cost. Of the nearly 2,000 men who had set out with Anson, only 188 made it home alive. Just three had been killed during the capture of the Spanish galleon.[62]

The true tragedy of scurvy at sea was that effective remedies were known from the appearance of the same symptoms within malnourished land-based populations. European and American folk medicine knew how to treat the early signs of the disease by eating cresses and spruce leaves,[63] which we now know are rich in vitamin C. Individual sea captains in the sixteenth century also discovered from experience that fresh vegetables and fruit, and especially citrus fruits, could cure scurvy. The Elizabethan explorer and privateer Sir Richard Hawkins noted in 1590 'that which I have seene most fruitful is sower [sour] oranges and lemons'.[64] But for centuries, navies and merchant ships adopted these remedies haphazardly, unable to preserve the foodstuffs for long periods at sea or failing to understand what the all-important ingredient in the diet was. We now call vitamin C ascorbic acid, because it was the long sought-after 'antiscorbutic' agent to prevent scurvy.

The Royal Navy in the eighteenth century was acutely aware of the problems of scurvy and tried to take measures to combat it. The problem was that the Sick and Hurt Board, which made recommendations to the admiralty on how Royal Navy ships should be provisioned, was dominated by classically trained physicians. The medical establishment at the time had a misguided understanding of the causes of scurvy, believing that it was an internal putrefaction of the body caused by poor digestion and the damp environment on ship. They dismissed as mere anecdote the first-hand accounts of sea captains and naval surgeons on which remedies were actually effective in combatting scurvy, as they did not fit with the prevailing theories of disease. Anson's voyage, for instance, had been provisioned with elixir of vitriol – sulphuric acid mixed with alcohol, sugar and spices – under the false belief that the antiscorbutic effect of citrus fruits was due to their acidity aiding digestion.[65]

But the appalling loss of life on Anson's disastrous voyage helped focus medical minds. In 1747, the naval surgeon James Lind compared the efficacy of several treatments that were considered at the time, including cider, vinegar, sulphuric acid and seawater, as well as oranges and lemons, in what is often described as the first controlled clinical trial. He took a group of sailors suffering from scurvy, ensuring that the cases were as similar as possible, and divided them randomly among the different treatment regimes. This systematic experiment demonstrated decisively that scurvy could be successfully treated with dietary supplements of citrus fruit – the sailors assigned to that treatment were able to return to duty within a week. He published his results, but unfortunately his treatise had little impact on naval practice.

The main problem was that although Lind's trial had shown the efficacy of citrus fruit in treating scurvy, he didn't appear to fully appreciate its significance: his publication recommended citrus fruit while still espousing a multitude of other, ineffective treatments.[66] What's more, he went on to develop a process for preserving citrus juice for long-term storage aboard ships by heating it to reduce the water content and concentrate the citrus juice, so that 24 oranges or lemons became just 100 millilitres or so of fluid. He claimed, without testing, that this fruit syrup would retain its antiscorbutic potency for years, when in fact heating destroys the vitamin C and makes the remedy ineffective.[67]

Scurvy continued to ravage not just mariners in the navy, but also those aboard merchant ships, including the slave ships crossing the Atlantic from Africa. Thomas Trotter was a surgeon aboard a slave ship based out of Liverpool in the 1780s, when he discovered that by straining fresh lemon juice and then bottling it under a thin layer of olive oil to help protect it from the air, the antiscorbutic properties of the juice were still effective more than a year later. Trotter, who would become an ardent critic of the slave trade, was convinced that this remedy could save 'the lives of poor retches who perished', but the slavers were unwilling to add extra victualling costs to their voyages.[68] Yet Trotter continued his campaign, eventually being appointed as the Physician of the Channel Fleet,[69] while other advocates of the use of citrus juice rose to influential positions on the Sick and Hurt Board.

It seems tragic that, although Lind's careful clinical trial had provided all the evidence that should have been needed, it wasn't for another forty years after the publication of his treatise that the Royal Navy began routinely provisioning its ships with lemon juice. The sea change had come with the reformation of the Sick and Hurt Board by physicians with practical experience of scurvy at sea who persuaded the admiralty to go against the theories held by the medical establishment. Experienced fleet admirals now also insisted on being provided with citrus juice. Rear Admiral Alan Gardner, for example, demanded daily lemon juice rations for his voyage to India in 1793 and arrived after a four-month stint at sea without a trace of scurvy among his crew. When the news reached home, other fleet commanders clamoured to be supplied with the effective antiscorbutic.

Finally, in 1796, the Royal Navy agreed to provision all its warships posted on foreign service with lemon juice, and from 1799, all those in home waters around the British coast received it in their rations as well.[70] The effects of this new policy were transformative. Between 1794 and 1813, the sick rate in the Royal Navy dropped from around 25 per cent to 9 per cent.[71] Scurvy had effectively vanished from British warships. Before the introduction of lemon juice rations, a ship could operate for only ten weeks before being overrun by scurvy. But after 1799, the Royal Navy was able to keep its squadrons and fleets at sea for four months without resupply of fresh provisions.[72]

The Royal Navy had finally conquered its greatest foe, granting it a decisive advantage over its rivals which were not yet fully safeguarding their crews. In particular, it meant the admiralty could now keep its crews healthy and its warships at full operational strength for the core component of British military strategy: the naval blockade.*

* Another biological limitation of the human body for long voyages is that we cannot drink seawater. This is not surprising. Even marine mammals that have evolved to inhabit the deep blue cannot survive on saline water and must extract body water from their prey (and also use the water molecules created during the metabolic breakdown of carbohydrates and fats). And their kidneys are designed to efficiently re-excrete salt in their urine. But humans have no such adaptations, and countless sailors have died of dehydration despite being surrounded by an ocean of water. As Samuel Taylor

NAVAL BLOCKADE

Many states through history have wielded naval power to further their interests. Since the early modern period, European states have deployed open ocean navies, the relative strengths of which waxed and waned over time in response to geopolitical threats or opportunities. But for the English, and then the British, a standing navy became a crucial force for defending the island against invasion. The Royal Navy, formed in the early sixteenth century, repeatedly clashed with the Spanish, French and Dutch fleets, evolving into a tool for not only safeguarding the homeland but also protecting Britain's growing maritime economy and overseas colonies.

The paramount role of the Royal Navy had always been to command the English Channel and shield the island against ships arriving from the continent, in particular the northern shores of France, Belgium and the Netherlands. But as England and then Great Britain increased its number of overseas possessions, the marine territory that needed to be defended grew significantly. Merchant shipping returning to Britain from the Caribbean and North America, or further afield from India and South East Asia across the northern Atlantic was funnelled through a key region of sea off the British coast known as the Western Approaches. The colonies themselves also needed protection, and if a rival European fleet was allowed to slip away into the vast expanse of the Atlantic, it could be lost without trace for weeks, until it appeared suddenly over the horizon to attack one of Britain's overseas colonies. Dedicating a sufficient number of warships to defend each of the valuable colonies, as

---

Coleridge had his ancient mariner lament, 'Water, water, every where, / Nor any drop to drink.' Mariners traversing that vast watery desert must take with them all the drinking water they will need, usually mixed with alcohol, as we have seen, to help keep it from spoiling. From the second half of the nineteenth century, however, distillation units for desalinating water became common aboard ships, finally enabling sailors to drink the sea.[73]

well as serving convoy duty along the sea routes and protecting the Channel, would be impossibly expensive, and dispersing the available ships too widely would make the navy ineffective in any of its roles.

How could the Royal Navy deploy its limited ships so as to adequately meet all the strategic demands placed upon it? The solution was the formation of the Western Squadron, a formidable concentration of warships stationed off the south-west coast of Britain. From here, the formation could move to achieve several goals at the same time: screening the entrance to the English Channel to shield against invasion; guarding the vulnerable home stretch of Britain's sea lanes in the Western Approaches to protect the lifeblood of its maritime economy; and, with periodic patrols down into the Bay of Biscay, preventing a rival power concentrating its naval forces which could threaten Britain's colonies. Dominance of the coastline and Atlantic waters of the western European promontory – those of the western shores of France and the Iberian Peninsula – became strategically vital in protecting British interests.[74]

By the 1740s, the Western Squadron had become the navy's main battle fleet and the lynchpin of Britain's naval shield.[75] It was commanded by Lord Anson after his long, scurvy-plagued campaign in Spanish America, who oversaw much of this growth.[76] The key strategy that enabled a single concentration of warships to be effective over such a large area was the implementation of a naval blockade.

It meant that during wartime, an enemy's principal ports could be sealed by a dominant naval force lying just out at sea, not only curtailing their military threat but also hindering their maritime trade and choking their economy. In ideal circumstances, an impregnable close blockade could be maintained to prevent enemy warships leaving harbour and combining into a strong fleet. But it is difficult to establish a permanent, complete guard over all major enemy ports to prevent their ships slipping away. So, more commonly, a loose blockade was employed, whereby small, manoeuvrable frigates kept a continuous close watch on the harbours. If the enemy ships dared to emerge, the main battle fleet stationed further away was alerted, via a chain of signal ships, and moved in to engage them with a superior force in decisive battle.[77]

Maintaining an effective blockade for many years at a time was an enormous logistical challenge, with stationed ships needing to be resupplied at sea and regularly relieved by replacement vessels from home ports, but the practice made efficient use of the ships at the admiralty's disposal.[78] As the late-nineteenth-century US naval strategist Alfred Thayer Mahan described it, 'Whatever the number of ships needed to watch those in an enemy's port, they are fewer by far than those that will be required to protect the scattered interests imperiled by an enemy's escape.'[79]

The Royal Navy was able to achieve command of the seas through this form of passive defence – controlling the waters of Western Europe not by annihilating the enemy's fleet (although the opportunity to do so would of course have been welcomed) but by confining their ships to harbour. Some ships or squadrons still had to be stationed to protect key strategic points elsewhere, but the principal defence of overseas possessions across the empire was enacted off the coast of Western Europe.

This strategic doctrine faltered, however, when France and Spain joined the War of American Independence. Despite being the largest navy in the world, the Royal Navy was stretched very thin and did not have enough ships to effectively blockade American ports, support its troops along the coast and protect the Caribbean colonies while also containing French and Spanish ships in their home ports. These quickly caused trouble for the British along the North American coast and in the Caribbean. And then there was the navy's old foe. Scurvy undeniably contributed to the loss of the Thirteen Colonies,[80] with British crews debilitated by the disease and the sailors that survived needing to spend long periods of convalescence in naval hospitals. Out of the 175,900 sailors recruited between 1774 and

1780, more than 18,500 died from disease, compared to only 1,243 killed in action. In 1782, at the height of the Revolutionary War, the sick list numbered 23,000 sailors out of a total force of around 100,000.[81] While British troops were harried on land by local strains of malaria (as we saw in Chapter 3), scurvy stalked Royal Navy sailors at sea.

By the time of the Napoleonic Wars twenty years later, however, reform of the Sick and Hurt Board had seen the Royal Navy provisioning all its warships with daily rations of lemon juice. With the numbers of crew falling prey to scurvy reduced dramatically, the Western Squadron successfully blockaded the main French naval bases at Le Havre in the English Channel, Brest at the north-western tip of France and Rochefort in the Bay of Biscay. The naval cordon around France was completed by the Mediterranean Squadron, commanded by Nelson between 1803 and 1805, which blockaded the southern coastal port of Toulon. Rochefort was held under a close blockade to keep the warships pinned and unable to leave harbour, whereas Nelson operated a looser blockade of Toulon, hoping to entice Napoleon's fleet based there to attempt a break for it so that he could force them into action and achieve a decisive victory at sea.

The Spanish fleet, based out of Cadiz, just north-east of the Straits of Gibraltar, and Ferrol, on the north-western tip of the Iberian Peninsula, was now allied with the French, and Napoleon planned to combine the two fleets and wrest control of the English Channel to provide safe passage for an invasion flotilla to land a French army on the southern English coast. French ships successfully evaded the blockade on several occasions but were unable to inflict a decisive victory over the British. The Royal Navy maintained their dominance of the waves, attacking the French at home and abroad at will, while British and allied shipping was free to trade around the world, helping to finance the war effort against imperial France.

But in January 1805, Napoleon's fleet was able to slip the net of the British blockade at Toulon and join forces with the

Spanish at Cadiz. They sailed to the Caribbean, pursued by Nelson, before turning north to pick up the westerly winds to carry them back to Europe. They intended to help break out the Brest fleet from their blockade and then sweep with their large numbers of ships-of-the-line into the English Channel to clear it of British warships. The plan was abandoned after French and Spanish losses at the Battle of Cape Finisterre, however, and the commander of the combined Franco-Iberian fleet, Pierre-Charles Villeneuve, decided to return to Cadiz. When Napoleon gave orders for the allied fleet to put to sea again and sail to Naples, Villeneuve made for the narrow gateway into the Mediterranean, the Strait of Gibraltar. Nelson's blockading fleet caught and engaged them 40 kilometres down the coast, just off the Cape of Trafalgar, on 21 October 1805.

Despite having spent many months at sea by this point, the men in Nelson's fleet were virtually free of scurvy.[82] The British Mediterranean fleet had suffered only 110 deaths from scurvy (and 141 hospitalisations) out of some 7,000 sailors between August 1803 and August 1805 during the blockade of Toulon. Nelson – who had almost died of scurvy himself in 1780 as a young captain during the American Revolutionary War[83] – had 'spent two years on the Victory, wanting ten days',[84] without once stepping foot onto dry land. In contrast, the French and Spanish ships were rife with scurvy – the Spanish commander at Trafalgar reported that some of his ships were weakened with more than 200 cases of scurvy each,[85] around a quarter of the men aboard a ship-of-the-line.

Not only were Nelson's crews shielded from the ravaging effects of scurvy, they were also more experienced at seamanship and better drilled in gunnery. Nelson played to these advantages with the unorthodox naval tactic of cutting through the line of enemy warships in two columns – a calculated gamble that paid off and secured the Royal Navy a decisive victory over the French and Spanish fleets.

The superior health of the British sailors, and the boost this gave to the effectiveness of sea power, was undoubtedly a major contributor to Nelson's fleet prevailing.[86] The conquest of scurvy and victory at Trafalgar had assured British naval supremacy, and the Royal Navy continued to dominate the seas until the Second World War. As naval historian Christopher Lloyd put it, 'Of all the means which defeated Napoleon, lemon juice and the carronade gun [a naval cannon] were the two most important.'[87] For almost another decade after the Battle of Trafalgar, the Royal Navy maintained a blockade of the French coast, choking France's wartime economy. Napoleon's response was the Continental System, imposing an embargo on British imports across those parts of Europe that were controlled by, or in coalition with, France. But Britain's control of the oceans meant it was able to divert its trade across the Atlantic to North and South America, and merchants also smuggled copious goods into Spain and Russia. This drove Napoleon to invade both Spain and Russia to enforce the boycott, resulting ultimately in his disastrous defeat and retreat from Moscow in 1812. Of the Grande Armée's 615,000 men who marched into Russia in 1812, only 110,000 frostbitten and half-starved survivors returned.[88]

The Royal Navy had finally vanquished one of its most enduring foes, the debilitating and lethal scourge of scurvy, while, for a time at least, it continued to ravage American and continental European sailors, who weren't provided with lemon juice.[89] The British had obtained a crucial strategic advantage, and while the practice of long-term blockade was supported by carefully planned logistics, biologically, their command of the seas was enabled by their effective countermeasures to our defective genome. Before long, however, lemons preserved in seawater came to be adopted by the navies of other nations,[90] and from the late 1800s, the advent of steam-powered ships greatly shortened oceanic voyages, which meant that few crews needed to spend months away from port, and the spectre of scurvy at sea receded. But for three and a half centuries, our defunct GULO gene played a pivotal role in the Age of Sail.

CITRUS AND THE RISE OF THE MAFIA

There's a little known side-story to this. The adoption of lemon juice rations by the Royal Navy created a huge demand for citrus: between 1795 and 1814, the admiralty issued some 7.3 million litres of lemon juice.[91] With the British climate not conducive to local production, sweet lemons had to be imported, mostly from around the Mediterranean. The admiralty first sourced its vital supply from Spain, but when the Spanish allied with Napoleon in 1796, it switched to buying Portuguese lemons, and then to those grown in Malta after the capture of the island in 1798.[92] Then, from 1803, Nelson turned Sicily into a giant lemon juice factory[93] to supply antiscorbutics for the Royal Navy worldwide.[94]*

Rural regions of the island, where the hot climate is ideal for citrus cultivation, have historically suffered from weak governmental control. Neither the Bourbon kings who ruled Sicily from 1735, nor the Savoy dynasty after Italian unification and independence in 1861, had enough political strength to enforce the law, including the rights of private property. The feudal system remained dominant in Sicily until the nineteenth century, with the barons exercising administrative, fiscal and judicial power over their dominions. After the auctioning of feudal lands in 1812, many tenants became landowners but were forced to hire private guards to protect their estates against the rampant brigands. Corruption and intimidation tactics prevailed.

---

* After 1860, the supply of citrus juice for the whole Royal Navy came from limes grown on British colonies in the West Indies.[95] In 1867, the passing of the Merchant Shipping Act required all ships of both the Royal Navy and the Merchant Navy to provide a daily ration of concentrated lime juice for their crews.[96] The nickname 'limey', referring to that ration, was first applied to British sailors, and then to British immigrants in the former British colonies (especially Australia, New Zealand and South Africa), before its usage in America extended to refer to all British people.[97]

The surge in demand for lemons driven by naval procurement created a commodity boom and injected a huge influx of cash into this unstable situation, greatly exacerbating the problems. It is out of this lawless landscape that the mafia emerged. The combination of high lemon prices, high levels of local rural poverty and a weak rule of law meant the owners of citrus orchards were extremely vulnerable to scrumping. Hundreds of pieces of fruit – representing a huge value on the market – could be stolen off the branches in a single night. With no centralised authority, the plantation owners were forced to turn elsewhere, hiring muscle to provide private security. This soon evolved into the practice of extortion and the threat of violence if the producer refused to pay the protection money. The mafiosi also acted as intermediaries between the rural growers and the international exporters in the harbours, so when a sale had been agreed, a lemon was placed on top of the gate to the grove to indicate that the property was under their protection. Their power only increased as the international demand for lemons snowballed: over just thirteen years between 1837 and 1850, exports rose from 740 barrels of lemon juice per year to over 20,000.[98]

By the 1870s, the organisation we would recognise as the modern Mafia had emerged, quickly expanding into wider racketeering and extortion and other organised crime. They came to infiltrate the economic and political institutions of the whole of Italy, and then the United States, which saw more than four million Italians, mostly from the impoverished south and Sicily, arrive between 1870 and the First World War.[99]

VITAMIN D AND A DEFICIENCIES

As a deficiency disease, scurvy afflicted mariners for hundreds of years during the Age of Sail. But other vitamin deficiencies have affected large numbers of people through history. Vitamin D, for example, is synthesised in our skin when it is exposed to the UV rays in sunlight. But in the far northern latitudes, not only do people need to wrap up against the cold, but the sunlight is weaker, and the winters are long and dark – so the human body is unable to produce enough vitamin D for itself. This vitamin is needed to help absorb calcium from our food, and so a deficiency leads to softened bones that become deformed. This manifests as the debilitating disorders of rickets in children and osteomalacia in adults.

Historical documents and skeletal remains indicate that rickets was common in northern parts of the Roman Empire and medieval Europe.[100] The indigenous Inuit across northernmost Canada and Greenland, and the Norse in Scandinavia, however, were largely protected as their diet contained sufficient amounts of oily fish such as cod and salmon that are rich in vitamin D.[101] And unlike vitamin C, vitamin D is fat-soluble and the body can store a sufficient reserve for months, so they could weather an inconstant supply.

Humans also need to obtain vitamin A from their diet, either in an active form from animals, or as an inactive 'provitamin' such as beta-carotene – the red-orange pigment in fruit and veg (including carrots, sweet potatoes, tomatoes and butternut squash) – some of which the body is then able to convert into vitamin A.[102] But many people around the world suffer vitamin A deficiency, especially those in developing nations dependent on white (milled) rice as a staple. Vitamin A deficiency affects around one third of children globally and is the leading cause of preventable childhood blindness; it also increases the risk of death from other common illnesses.[103] One solution to this widespread problem is the genetic modification of rice crops to produce beta-carotene in the edible parts of the grain, creating 'Golden Rice'.[104] Technology is being used to add genes to one of our most ancient staple crops in order to compensate for biochemical inadequacies in our own make-up.

We'll turn now from the historical ramifications of defects in our genetic code, to the numerous processing glitches and biases in our psychology.

Chapter 8

Cognitive Biases

It has been said that man is a rational animal. All my life I have been searching for evidence which could support this.
—Bertrand Russell

On the evening of 3 August 1492, Christopher Columbus set sail from Palos de la Frontera in southern Spain. His small flotilla was made up of a large carrack named the *Santa María*, his flagship, and two smaller caravels. They stopped first in the Canary Islands to take on board provisions and make some repairs, before turning due west across the wide expanse of the Atlantic. This voyage was to become historic, but not for the reasons Columbus anticipated.

Columbus believed that the route west across the sea to the Orient was shorter than the overland journey east across Asia. He based this on his calculations of the circumference of the Earth and the accounts of explorers who had traversed Eurasia along the Silk Roads. Working from a map sent to him by Paolo dal Pozzo Toscanelli, an Italian astronomer and cartographer, Columbus also believed Japan to lie much further out from the East Asian coast, so it could be used as a stop-off to reprovision his ships after the long sea crossing.

We now know that the numbers Columbus used had simultaneously underestimated the circumference of the Earth and overestimated the breadth of the Eurasian landmass; he also assumed that Japan was much closer to Europe than it really was. By Columbus's reckoning, there was certainly no room for an unknown continent lying between Europe and China. As far as he was concerned, he wasn't embarking on a voyage of discovery but undertaking a calculated attempt to reach a known destination by a new route.[1]

But after a month of sailing west, his lookouts still hadn't seen any sign of land, and the crew were starting to get unsettled. Columbus knew that he could probably only keep his men in line for a short while longer before he would be forced to turn back or risk mutiny. He figured they must have somehow missed Japan but were surely nearing the coastline of China. Finally, in the early hours of 12 October, after five weeks at sea, they sighted land.

The tale of Columbus's discovery of the Americas is a familiar one. But what's less well known are the extraordinary mental gymnastics the explorer performed to maintain his belief that his voyage had actually reached the Orient and not a strange, new land (he had in fact landed in the Caribbean). It's a clear example of a particular cognitive glitch that can affect us all.

There were clear signs on this first voyage that should have alerted Columbus to the fact that he hadn't reached East Asia. The interpreter he had brought with him spoke several Asian languages, but he couldn't make himself understood by the inhabitants of the islands they encountered. Instead of the civilised and cultured people described by Marco Polo, the Venetian merchant who had travelled to China overland in the late thirteenth century, the local populations strolled around naked in a seemingly primitive existence. And once they got over the language barrier, they realised that nobody knew of the powerful and magnificent khan who was supposed to rule China.

Columbus and his men also failed to find any of the valuable spices of the East: cinnamon, pepper, nutmeg, mace, ginger and cardamom. The first day after making landfall, his sailors observed the locals carrying around bundles of dry, brown material, which they initially believed to be the curled bark of the cinnamon tree. But on closer inspection it turned out to be dried leaves. And even more strangely, the locals had the habit of burning these leaves and inhaling the smoke: as we have seen, the sailors were introduced to the habit of smoking – a practice never reported from the Orient. Columbus was never able to locate any cinnamon trees in his exploration of the new lands.

He was also unable to find pepper. The locals did make flavoursome stews with an added spice, but Columbus realised this 'pepper' had a very different appearance and taste to that which he had hoped to bring back from China. These New World peppers belonged to the Capsicum family, which includes chilli and cayenne peppers, bell peppers, pimento, paprika and tabasco. Other plants were misidentified by Columbus's crew. The gumbo-limbo tree was mistaken for the mastic tree, which oozes a valuable resin from cuts made in the bark. Another plant was determined to be medicinal Chinese rhubarb. They also found agave, a rosette of thick, fleshy leaves with spines, and mistook it for aloe.*[2]

Columbus made a total of four voyages west across the Atlantic, extensively exploring the coastlines of Cuba and Hispaniola, as well as many of the smaller Caribbean islands and parts of the coast of the Central and South American mainland. But over these explorations, stretching over twelve years, with all their unexpected encounters and findings, he never accepted that he hadn't reached the Orient at all, but another place

---

* At the time, the yellow sap from aloe was dried and taken in small doses as a potent laxative; today, we use it in skin moisturisers and cosmetics. In this case, though, Columbus might be excused his misapprehension. While the American agave plant is distinctly larger than aloe – which is native to Madagascar, the Arabian Peninsula and islands of the Indian Ocean – the two do have the same general appearance. However, they aren't closely related at all; instead, they resemble each other because they have developed similar biological solutions to the same survival challenges – a process known as convergent evolution. Both agave and aloe developed thick, fleshly leaves to store water, and spines to protect this vital cache against herbivorous animals, to survive in hot, dry climates.

entirely. Very little in this new land matched up with the reports from travellers to East Asia. But Columbus viewed everything he saw in the new world through the prism of his own deeply held beliefs. He clung to any scrap of evidence that seemed to support his prior expectations, while any counter-indications that challenged these preconceptions – and there were many – were reinterpreted, downplayed or simply ignored.

Columbus suffered what we would today call 'confirmation bias'. This is our tendency to interpret new information as further confirmation of what we already believe while disregarding evidence that challenges our beliefs.[3] This tendency to resist re-evaluating or changing our beliefs once they are established has been recognised as a human trait for centuries.[4] As Sir Francis Bacon wrote in 1620: 'The human understanding when it has once adopted an opinion ... draws all things else to support and agree with it. And though there be a greater number and weight of instances to be found on the other side, yet these it either neglects and despises, or else by some distinction sets aside and rejects, in order that by this great and pernicious pre-determination the authority of its former conclusions may remain inviolate.'

Five hundred years after Columbus, confirmation bias was also a powerful influence in the production of the intelligence report that concluded Saddam Hussein was building weapons of mass destruction, the *casus belli* used to justify the US-led invasion of Iraq in 2003. Iraq did not in fact possess any WMDs, and subsequent investigation into how the case had been prepared revealed that the analysts' conviction had been so firm that they viewed all information through the prism of their prevailing paradigm. Instead of weighing up pieces of evidence independently, they readily accepted supporting information and simply disregarded contradictory signs, without ever properly questioning their initial hypothesis. For example, a few thousand aluminium tubes purchased by Iraq and intercepted by the security forces in Jordan were determined to be intended for gas centrifuges, equipment used to enrich uranium to weapons-grade. But although the tubes *could* be employed for such a purpose, the evidence was that in fact they were better suited for conventional uses, such as building rockets, which was discounted. Confirmation bias merely fortified an unsupported presumption and led to war.[5]

Confirmation bias is also the reason why two people with opposite views on a particular subject or issue can be shown the exact same body of evidence and both come away convinced they've been vindicated by it. No matter how balanced and neutral the new information may be, our tendency to cherry-pick the details that confirm our own preconceptions will leave us thinking our own position has been validated. Ironically, it's the sort of information-processing bias that we can be utterly blind to in ourselves and yet find glaringly obvious and infuriating in others.[6]

This glitch in our cognitive software is a major cause of the increasingly polarised politics that we find in Britain, the US and the rest of the world today.[7] And it's a vicious cycle: once someone believes that the other side, or a particular media outlet, cannot be trusted and is not presenting a balanced or objective viewpoint, it becomes all the easier to discredit or dismiss their points and arguments.

The problem of cognitive bias is further exacerbated by the search engines and social media platforms from which many of us today source our news and information about the wider world.[8] The algorithms underlying these services are designed to analyse what each individual user has clicked, liked, favourited, searched for or commented on in the past, and then start feeding them the content they are deemed most likely to engage with (and share with their own network of online contacts). In this way, our online world becomes an echo chamber feeding us more of the same and in doing so further intensifying any confirmation bias. As a result, many of us nowadays are no longer exposed to contrary ideas or political views. The web is fragmenting into myriad personalised bubbles – it has become the splinternet.

MENTAL GLITCHES

The human brain is an absolute marvel. It excels in computation, pattern recognition, deductive reasoning, calculation and information storage and retrieval. On the whole, it is far more capable than any computer system or artificial intelligence we've ever been able to construct. Indeed, considering that our cognitive abilities evolved to keep our Palaeolithic ancestors alive on the plains of Africa, the versatility of our brains to handle mathematics and philosophy, compose symphonies and design space shuttles is all the more staggering.

Despite all these incredible feats, the human brain is far from perfect. While it often performs admirably in making sense of a complex, chaotic world, it can also fail spectacularly. Shakespeare has Hamlet declare that humans are 'noble in reason . . . infinite in faculty', but this patently isn't true. Our brain has clear limitations in terms of its maximum operating speed and capacity.[9] Our working memory, for example, can only hold around three to five items (such as words or numbers) at once.[10] We also make mistakes and bad judgement calls, particularly if we're tired, our attention is overtaxed or we're distracted. Limits, however, are not defects:[11] any system must operate within its own constraints.

But in many cases people make the same mistake in the same situation; the error is systematic and predictable, so researchers can set up a scenario to reliably reproduce it.[12] That seems to hint at something deeper – a fundamental part of the brain's software revealing itself through consistent glitches.

These deviations from how we would expect a perfectly logical brain to function are known as cognitive biases. We've explored how confirmation bias led Columbus astray, but there are many other kinds. These cognitive biases can be broadly organised into three categories: those affecting our beliefs, decision-making and behaviour; those influencing our social interactions and prejudices; and those distorting our memories.

They include the 'anchoring effect', our natural tendency to base an estimation or decision too heavily on the first piece of information we're given. This is why the opening figure in a marketplace haggle or a salary discussion can be hugely influential on the outcome. 'Availability bias' is the tendency to ascribe greater importance to examples that are easily remembered. For instance, you might feel that flying by plane is more dangerous than travelling by car. But while a catastrophic plane crash may kill a great number of people all at once and is more likely to be reported on the news, the cold hard stats clearly show that, mile for mile, you are a hundred times more likely to be killed in a car than a plane. The 'halo effect' is the tendency for our positive impressions of one aspect of a person to be generalised to other unrelated traits of that individual. Then there's the well-known 'herd bias', the proclivity to adopt the beliefs or follow the behaviour of the majority so as to avoid conflict. 'Rosy retrospection' is the tendency to recall past events as being more positive than they actually were. 'Stereotyping' or 'generalisation bias' describes the expectation that an individual member of a group will have certain attributes believed to be representative of that group. And so on. In fact, if you look-up the 'List of Cognitive Biases' on Wikipedia, you'll see over 250 named biases, effects, errors, illusions and fallacies – though many of these are so similar they probably reveal different facets of the same underlying cognitive process.

While not everybody is equally susceptible to all biases, it is certainly true that everyone is affected by them in some way or other. Indeed, an individual's failure to recognise that they are influenced by bias is a bias itself, known as the 'bias blind spot'. What's more, it turns out that even being aware of the existence of a bias is not enough to vaccinate our minds against its effects. Cognitive biases are systemic, intrinsic parts of how our brain works and they can be extremely difficult to counteract.*

* People with autism spectrum disorder (ASD) often show 'enhanced rationality' (compared to neurotypical individuals) in seemingly being less

It is difficult to dispute, therefore, that our mental operating system is beset with a litany of bugs and glitches. What is more controversial, however, is exactly what causes these biases. Why is it that the human brain can deviate so much from what you might expect the rational or logical response to be, and in such predictable ways?

Many of our cognitive biases appear to be the result of the human brain attempting to function as well as it can with its limited computational capabilities, using simplified rules of thumb known as heuristics. These are efficient cognitive short-cuts, a set of time-saving tricks, that allow us to make rapid decisions without fully processing every piece of available data when time is limited or information is incomplete.[14] They are both fast (based on simple processes) and frugal (using little information), so we effectively swap a difficult question for an easier one.[15] In everyday situations, it is almost always better to make a timely judgement that is adequate than spend ages gathering information and deliberating deeply to try to come to a decision that is absolutely optimised. The perfect is the enemy of the good-enough, and this is particularly true when your very survival may be at stake: a late decision may end up being the last one you ever make. While heuristics serve us well most of the time – at least often enough in our evolutionary history that they've helped us survive – they are prone to trip us up in certain circumstances.

Some biases that evolved in our ancestors' natural environment cause inappropriate responses in the modern world. Take the 'gambler's fallacy', for example. This cognitive bias makes us believe that a random event is more likely to happen soon if

susceptible to cognitive biases and more objective when it comes to reasoning, judgements and decision-making. In particular, people with autism appear to rely less on intuition, be less swayed by irrelevant information and show less aversion to considering negative information. If not all cognitive biases are universal, this enhanced rationality in ASD offers a promising opportunity to study the mechanisms in the brain behind both rational and irrational thinking.[13]

it hasn't happened for a while; or conversely, that an event that has just occurred is less likely to happen again immediately. It's the intuitive feeling, for example, that if a coin toss has just landed heads ten times in a row, it's much more likely to come up tails next.

There's a famous example. On 18 August 1913, in the Monte Carlo casino in Monaco, black came up on the roulette wheel an incredible 26 times in succession. There was a scramble of gamblers trying to get their bets down on red, betting higher and higher stakes in their mounting certainty that the run of blacks simply must stop. But they lost time and time again, and the casino made millions.[16] The roulette wheel wasn't rigged, and in logical terms, each spin had exactly the same probability of black coming up again (48.7 per cent on a European roulette wheel with one 'o'), regardless of how many times it had come up previously.

Of course, neither a coin nor a roulette wheel has any memory of past events, or indeed any ability to control the outcome of the next toss or spin – each is completely independent of those before. We know this must be true if we think about it logically, but that doesn't stop the pervasive thought in the back of our mind that our number surely *must* be due to come up any moment now, to balance things out.

Likewise, there is the 'hot hand fallacy'; originating in basketball, this is the belief that a sportsperson who has been successful in scoring a basket a few times already will have a greater chance of success in following attempts. Beyond sports, this belief is also held by a gambler convinced they're on a roll and should keep going to capitalise on their lucky streak before it runs out. Of course, in sports there are reasons why a talented player might be performing slightly better than average, perhaps through gaining confidence after successes in a game. But in most circumstances when it feels like you're on a hot streak, it's no more than chance – like heads coming up ten times in a row every now and then.

Both the gambler's fallacy and the hot hand fallacy have their roots in the same cognitive premise. This is the assumption that similar events are not independent of each other – that there is a connection between separate rolls of the dice, spins of the roulette wheel or shots at the basket. Such an assumption is fallacious in settings like a casino, where extreme care has been taken – the slot machines calibrated, the roulette wheel balanced, the card deck thoroughly shuffled – to ensure that the outcome of the next event is totally unconnected to previous ones. But such truly random distributions were pretty rare in our ancestral environment, in which our cognition evolved to perform.

The natural world is full of patterns. For example, many resources of value to our hunter-gatherer ancestors were not distributed randomly but could be found clustered together: berries on trees, particular plants flourishing in the same habitat, game animals congregating in herds. In these natural circumstances, if you find something you're looking for, you are indeed likely to find more in the same place.

Thus some cognitive biases aren't due to some underlying faulty wiring in our brains; they are design features of a cognitive process that has been honed by evolution to fit the characteristics of our natural environment.[17] Or you could say that such behaviour is not *illogical* but *ecological*.[18] It is the artificiality of a casino or experimental set-up in a psychology lab, where randomness has been implemented, that creates an apparent bug in our coding.

Studies of cognitive biases have led researchers to propose two different processing systems in our brains. The first is intuitive and quick, runs subconsciously and employs heuristics for rough and ready responses; while the second is only activated when the task requires it, operates on the output of the first, is analytical and slow and requires concentration. The second system evolved more recently – it is a newer layer of software layered on top of more ancient cognitive infrastructure. The problem is that the first, autonomous system cannot be turned off, and as the second system demands mental exertion, when we're tired or pressurised we default to the easy, sometimes irrational intuitions of the first.*

THE CURSE OF KNOWLEDGE

Humans have evolved a remarkable ability to consider the world from the perspective of another person. This lets us better understand another individual's motivations and intentions, and so equips us to predict, or even manipulate, their subsequent behaviour. Possessing such a 'theory of mind' is a crucial function for succeeding in social interactions and a key step in childhood development. Part of this capability is understanding that another person can have access to different information from yourself, and thus have different beliefs about the world, including some that you know not to be true.

This has been investigated in young children with what is known as a 'false-belief test'. In the experiment, a scenario is acted out in front of the child that involves a basket, a box, a chocolate and two dolls, Sally and Anne. The doll called Sally places a chocolate in the basket before leaving the room. Anne then takes the chocolate from the basket and puts it in the box. When Sally comes back into the room, the infant is asked where she will look for her chocolate. At around four years old, a child has developed a good theory of mind and recognises that Sally's information and thus beliefs are different from the child's own, and that she will try to find the chocolate in the basket even though the child knows it is now in the box.

Despite our exceptional theory of mind, we suffer from a variety of egocentric cognitive biases – tendencies to assume that the knowledge or beliefs held by other people are like our own. One such egocentric bias is known as the curse of knowledge.[20] We

---

* Readers may already be familiar with these two types of cognitive systems from Daniel Kahneman's popular book *Thinking, Fast and Slow*, and over the years, over twenty other 'dual process theories' have been proposed.[19]

are persistently biased by our own knowledge and experience when trying to consider how others, with access to different or less information, would interpret a situation and then act. Once we know something, we have trouble conceptualising what it's like not to know it. In the run-up to the Iraq War, US Defence Secretary Donald Rumsfeld in 2002 used a now-infamous categorisation of understanding, distinguishing between *known knowns, known unknowns* and *unknown unknowns*. The curse of knowledge, however, operates in a fourth domain of *unknown knowns* – when we don't realise what we know that others don't.

We can all recall getting frustrated trying to explain something to a partner or friend who repeatedly fails to understand what we mean when, as it turns out, our description omitted a key bit of information that we'd assumed was obvious. It crops up frequently in everyday life. But in high-stakes situations, the consequences of this form of cognitive bias can be catastrophic.

One of the most infamous military blunders in history was the Charge of the Light Brigade in 1854. The Crimean War saw Britain, France, Sardinia and the Ottoman Empire ally against Russia, ostensibly to counter its expansion into the Balkans. During the siege of Sevastopol, Russian's main naval base on the Black Sea, the British Army found itself on the defensive in the nearby port town of Balaclava.

On the morning of 25 October, Russian infantry had overwhelmed three British earthwork forts, or redoubts, positioned along the Causeway Heights, a low range of hillocks that separated two valleys just to the north of Balaclava. The loss of these positions left that vital port and British supply lines dangerously exposed. The British commander in Crimea, Lord Raglan, therefore ordered his cavalry to retake these redoubts and stop the Russians removing the British cannons positioned there. But peering through his telescope, to his abject dismay Raglan watched his Light Brigade of Cavalry instead ride down the length of the north valley and, in a seemingly suicidal frontal assault, charge straight into the heavily defended Russian artillery battery at the end. Man and beast alike were torn apart by the cannons firing point-blank, and after reaching the Russian position, the shredded squadrons were immediately forced to turn around and retreat back down the valley. Within a matter of minutes, over half of the 676 cavalrymen had been killed or wounded, and almost 400 mounts littered the valley floor.

The foolhardy bravery of the cavalrymen riding into the Valley of Death was immortalised at the time by the poet laureate, Alfred, Lord Tennyson, in his eponymous poem, 'The Charge of the Light Brigade':

> 'Forward, the Light Brigade!
> Was there a man dismayed?
> Not though the soldier knew
> Someone had blundered.'

There had clearly been a disastrous breakdown in battlefield communication, but who exactly had blundered? The episode is often held up as an example of staggering senior ineptitude – but could there have been something deeper behind the tragedy?

Eyewitness reports paint a vivid tapestry of the events leading up to the disaster and reveal how the orders could have become so confused. Raglan issued his first order of the day to the cavalry at about 8 o'clock in the morning: 'Cavalry to take ground to the left of the second line of redoubts occupied by the Turks.' This was ambiguous and confusing. There wasn't a second line of redoubts, and the direction 'left' is entirely dependent on the perspective of the viewer. Whose left did he mean? On this occasion, the commander of the British cavalry, Lieutenant General George Bingham, 3rd Earl of Lucan, interpreted correctly and moved his Heavy and Light Brigades accordingly.

Later in the ongoing battle, however, the commands became even more confusing. At 10 o'clock, Lord Raglan ordered his cavalry to advance and take advantage of any opportunity to recover the Heights, using the support of the infantry who were already en route. Raglan believed he had ordered an immediate advance, but Lord Lucan understood that he should wait for the infantry before moving forwards. It was at this point that, peering through his telescope, Raglan saw that the Russians were preparing to remove the British guns from the captured redoubts. Wishing to stress the urgency of his previous instruction, Raglan dictated the calamitous order: 'Lord Raglan wishes the cavalry to advance rapidly to the front – follow the enemy and try to prevent the enemy from carrying away the guns.' The written message was given to Captain Nolan, an aide-de-camp and a fine horse-rider, to be conveyed to the cavalry down on the floor of the northern valley. This choice of messenger was unfortunate, as Nolan was hot-tempered and contemptuous of Lord Lucan and his deputy, Lord Cardigan, both of whom he considered dithering aristocrats lacking courage.*

When Nolan handed over the written order, Lucan read it with bafflement. The wording was ambiguous and Lucan expressed his confusion over exactly where he was being directed to advance.

'Lord Raglan's orders,' Nolan retorted curtly, 'are that the cavalry should attack immediately.'

'Attack, sir!' exclaimed Lucan. 'Attack what? What guns, sir?'

Wafting his hand vaguely along the valley Nolan replied disdainfully 'There, my lord, is your enemy! There are your guns!'

It seems likely that here Nolan had injected his own interpretation – the actual order made no mention of attacking. Raglan's intentions were most probably only for the cavalry to move towards the redoubts to pressure the Russians to abandon the positions and be forced to leave behind the British guns.

Eyewitnesses report that Lord Lucan stared sternly back at the disrespectful Captain Nolan but didn't pursue his querying of Raglan's orders any further. The guns he could see from his

* Two different kinds of warm clothing take their names from this particular battle in the Crimean peninsula in 1854. The balaclava headwear is named for the port town near Sevastopol, after British troops took to wearing them to keep warm in the bitter cold. And the officer who led the Charge of the Light Brigade, James Brudenell, 7th Earl of Cardigan, is remembered for his open-fronted knitted top.

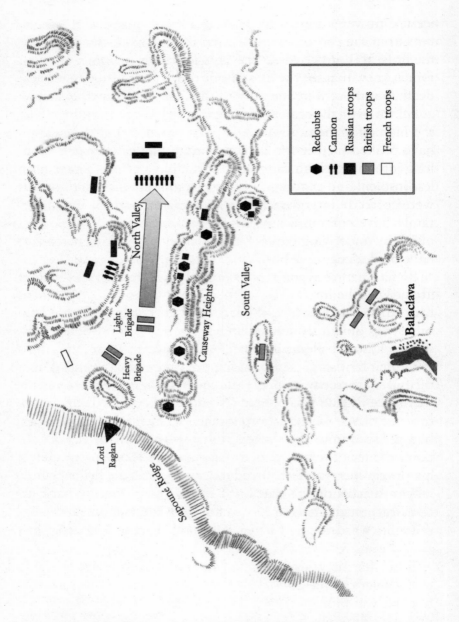

Map showing the Charge of the Light Brigade, the redoubts along the Causeway Heights, and Lord Raglan's elevated viewpoint from atop the Sapoune Ridge.

viewpoint were those of the Russian artillery position glinting at him from the end of the valley. After some hesitation he issued the order to Lord Cardigan to charge the enemy. The light cavalrymen knew that they would be riding out to meet almost certain death, but those were their orders and they had to be obeyed.

All three of the main characters involved carry a portion of the blame for the mortal disaster that ensued. Lucan ought to have pushed further for clarification instead of charging headlong into the Russian cannon position. Nolan should have been less insolent and goading, ensuring that his commander's orders were correctly interpreted. And Raglan, for his part, could certainly have communicated his orders more precisely. But the ultimate, underlying factor that spelled doom that day seems to have been a cognitive bias.

Raglan hadn't even considered there could be any ambiguity about what he intended. He knew what he meant. How could his clear orders possibly be misconstrued? In his mind, the final order was obviously a follow-on from the previous one: Lucan was to advance along the Heights and recapture the redoubts with their British guns. But Lucan had no way of knowing that this wasn't a separate order designating a new objective. Furthermore, Raglan was positioned atop the Sapoune Ridge at the top end of the valley, and from this elevated vantage point he had a panoramic view across the entire battlefield. From his point of view (literally!), the intended targets were obvious and his orders were perfectly clear. But what Raglan had failed to take into account was that on the valley floor, Lucan's perspective was much more limited. Lucan's view of the captured redoubts was obscured by the hilly landscape, and he couldn't see the guns Raglan was referring to.

While his intentions were clear to himself, Raglan had failed to consider that he and Lucan had access to different information. He was blind to the fact that his orders could be interpreted in a very different way. The Light Brigade was slaughtered by the curse of knowledge.[21]

The Charge of the Light Brigade is a classic example of what has come to be known as a common ground breakdown. This term describes how two people trying to communicate and coordinate their actions can, without realising it, be working with different information and thereby lose their common ground. In this case, Raglan, suffering from the curse of knowledge, failed to consider that Lucan may not have had access to the same information and battlefield views as him. Indeed, Lucan didn't have the crucial knowledge that the Russians were preparing to remove the British guns from the captured redoubts but also didn't realise that he was supposed to have that knowledge – that he might be missing some information. So when Lucan didn't request that information from Raglan, this in turn led Raglan to conclude that Lucan already knew. Out of this initial curse of knowledge, a whole cascade of incorrect inferences followed. The two men failed to realise that there was a fundamental mismatch in their beliefs. From Raglan's point of view, Lucan's subsequent actions were utterly baffling.[22]

The Charge of the Light Brigade was a notorious military disaster borne of the curse of knowledge. But the underlying cognitive bias doesn't just influence our assessment of the beliefs held by other minds; it also clouds how we consider our own past beliefs. Often, insights based on new information make us unable to appreciate why we previously believed something different. This distortion of our memory also manifests as the 'hindsight bias'. This is the tendency to perceive a past event as having been more predictable than in fact it was; this, in turn, can make us overconfident in our ability to predict the outcome of future events. This cognitive bias plays a crucial role in a particular literary genre.

Most of us enjoy a murder mystery or whodunnit. It is a popular form of storytelling where the narrative slowly builds up to a big denouement when the unexpected identity of the killer is revealed. There is often a surprising twist delivering a key piece of information about an event or character that seismically shifts our perspective on the entire story. The conclusion is satisfying, and in retrospect all the signs were there clearly enough.

But this narrative device presents a deep problem for the writer. How do you engineer a twist that is genuinely surprising to the audience, but which also fits naturally with the information already presented so that it doesn't appear as a fudge or cheat? The challenge faced is to sprinkle throughout the narrative enough clues whose true significance only becomes apparent at the conclusion – seeds that have to be planted firmly enough in the reader's mind that they are remembered when the denouement is delivered at the end, but which are subtle enough so not to alert the reader prematurely and give away the solution to the mystery.

This is where the curse of knowledge comes in. Once this additional, perspective-shifting piece of information is revealed, and the previous clues are reinterpreted in a new light, this cognitive bias gives you the impression, in hindsight, that the ending was possible to predict all along – indeed, that it was all glaringly obvious when you come to think about it. Knowing what you know now, the signs were all there in plain sight and you can't imagine how you missed them previously.

This cognitive bias is exploited masterfully by mystery writers like Agatha Christie, whose *The Murder of Roger Ackroyd*, published in 1926, is widely considered to be one of the most influential crime novels ever written. The plot involves Hercule Poirot investigating the murder of a friend with a new assistant, who serves as the narrator of the story. A breathtaking plot twist is delivered in the final chapter, like a rug pulled from beneath the reader's feet, when it is revealed that – spoiler alert! – the assistant, who as the narrator the reader has grown to trust, had in fact been the murderer. The same narrative device is used by Vladimir Nabokov in *Pale Fire* (1962) and Jim Thompson in *Pop. 1280* (1964). In cinema notable examples include the dawning realisation of the actual identity of criminal mastermind Keyser Söze in the film *Usual Suspects* (1995), and the twist endings delivered by director M. Night

Shyamalan. In each case, all the clues had been there, hidden in plain sight by the clever misdirection of the writer, but the curse of knowledge taints our retrospective perception to make them appear as if they had been obvious all along.[23]

CONCORDE DEVELOPMENT

Cognitive biases affect each of us individually, but they also have a powerful influence on decisions made collectively.

In some circumstances, judgements made by a group of people are far more accurate than those made by any one of its individual members. For example, in a common game played at summer fairs, a giant glass jar sits atop a counter, crammed with brightly coloured small sweets, challenging you to correctly guess how many there are. The winning guess will of course be an almost complete fluke. It's impossible to actually count all of the sweets inside the jar – you can only see those ones pressed against the transparent sides – and although you could make a fairly good, educated guess, this will only get you into the right range. But if you're lucky, no one else will have guessed any closer than you.

What's really interesting, though, is that if you look closely at all the guesses that were submitted, there'll be a broad spread: most in the right ballpark, with others off by a fair bit and a handful that are wildly off. This sort of distribution – known as the bell curve – turns up all over the place: from heights or IQ scores in a national population to the number of frogspawn laid by each female frog in a pond or the probability of getting a given number of heads when tossing a coin hundreds of times. In the case of guessing the number of sweets, the distribution is due to statistical variance between individual estimates, but the key point is that these fluctuations or errors are essentially random, and so over a large number of guessers, the higher and lower estimates tend to cancel each other out. If you take the median of the whole distribution of guesses – the value that lies exactly halfway down the list after you've sorted all the guesses into numerical order – you often end up with a number astonishingly close to the actual answer.* This is known as the wisdom of the crowd. Any given individual probably does pretty badly, but in a large pool of independent guessers, the scatter of errors converges towards a remarkably accurate answer.

On that basis, you might anticipate that other judgements and decisions made by a group are also more rational and less affected by cognitive biases. However, the problem with the biases inherent in all our brains is that they are not random: they are systematic. As they work in the same way for each of us, an aggregate will not cancel out the errors. Indeed, if anything, cognitive biases can serve to amplify them.[24]

In the late 1950s, aeronautical engineers in both France and the UK began working separately on plans for a bold and completely new kind of aircraft: a supersonic passenger jet.

It's hard to overemphasise just how staggeringly ambitious this endeavour was at the time. The first passenger jetliners had only entered service at the beginning of the decade, and the only aircraft at that time capable of exceeding the speed of sound were small fighter jets with one or two crew members. Even they could typically only fly supersonic for very short periods, before these enormously fuel-thirsty machines needed to land again. And along came this hare-brained idea to build a commercial jet for 100 or more passengers in armchair and champagne comfort, cruising at supersonic speeds of up to Mach 2 – twice the speed of sound, or around 2,200 kilometres per hour – continuously for over three hours. Both Britain and

---

* You need to use the median rather than the arithmetic mean since this latter kind of average can be skewed by extreme outliers – people that guessed 'one million sweets', say, or 'no sweets at all' just for a laugh. In a symmetrical distribution, known mathematically as a 'normal' distribution, the shape of the graph is a perfect bell curve, and the mean, median and mode (the most common number, appearing as the peak of the hump-shaped graph) all fall at the same number, smack in the middle.

France began developing their own supersonic transport plane, but these two parallel projects were combined in 1962 when a bilateral agreement – a *concord* – was signed between the two countries. But even from the early stages, the development of this bold new supersonic aircraft was plagued with problems and setbacks.

The technical challenges themselves were formidable. The very fact that it was so very unconventional meant that much of this new Concorde aircraft had to be designed from scratch. The wings needed to generate lift for stable flight at around 2,200 kilometres per hour as well as at 300 kilometres per hour for landing. The engines needed to be fuel efficient enough so that the aircraft could fly for hours. The plane needed a computer-controlled system to continually adjust the flow of supersonic air through the engine intakes and a much more comprehensive and sophisticated autopilot. Flying that fast, the outer layers of the aircraft would heat up to about 100 °C, which meant that the metal airframe needed to be able to accommodate expansion of up to 30 centimetres. Engineers also had to ensure that the fuel tanks didn't spring leaks with all the flexing of materials, and hefty air conditioners were needed to protect passengers and crew from the heat. Concorde was fitted with afterburners to help give it the extra shove needed for take-off and to push it through the sound barrier. At the low speeds needed for approach to landing, the aircraft needed to significantly pitch up, and so in order for the pilots to be actually able to see the runway, the entire sharply pointed nose was mechanically rotated to droop down. And all this needed to be designed and tested, using scale models in wind tunnels, at a time long before detailed computer simulations were possible.

While all these technical hurdles were causing delays and cost overruns, the financial viability of the aircraft – if it could ever be completed – was also being cast into doubt. Due to the nuisance of sonic booms, in 1973, the US pre-emptively banned supersonic commercial flights overland,[25] which scuppered the Concorde's hopes of running fast services between American cities on the east and west coasts. It became clear that this new aircraft would be restricted to routes over oceans or other unpopulated regions such as deserts. Last but not least, the rising costs also called into question which airlines might actually want to purchase the aircraft when it was completed.[26]

By the time the Concorde project was finally completed in 1976, fourteen years after the signing of the Anglo-French development agreement, the costs had ballooned. The original cost estimate in the early 1960s had been £70 million (equivalent to £1.42 billion today) but when the aircraft was eventually delivered, the total cost of the program had swollen to around £2 billion[27] (equivalent to £13 billion today).[28] The costs of research, development and production could never be recovered.[29] Only fourteen Concorde aircraft were ever sold, and only to the state airlines of the two countries involved – Air France and British Airways – with their governments agreeing to cover the support costs of the aircraft.[30]* Concorde was a marvel of aeronautical engineering but an economic failure for the parties involved in its long and expensive development.

It's not unusual for R&D programmes or construction projects to experience unexpected delays or cost overruns, but in any endeavour that meets substantial problems there is the recurring question of whether it is worth pushing on or better to cut losses and terminate the project. In this case, surely, once the extraordinary technical challenges of supersonic travel and the burgeoning development costs had become apparent, and when the market for the new aircraft was shrinking, the Concorde should have been abandoned as a commercial venture.

---

* Concorde was not in fact the first commercial supersonic aircraft. That honour falls to the Soviet Tupolev Tu-144 (dubbed 'Concordski' in the West), which took off on its maiden flight just two months before the British-French Concorde. But the Tu-144 suffered from reliability and safety issues and flew only 55 passenger flights in the three years of its operation.[31] The US had also been designing its own supersonic passenger airliner, the Boeing 2707, but its development was cancelled in 1971 because of rising costs and the lack of a clear market[32] – the same issues that plagued Concorde.

But the two governments financing the project decided to continue because they had 'too much invested to quit'.[33] And, of course, Concorde had become an issue of national pride and prestige, and the Anglo-French agreement meant that once the ball had started rolling, the project was much harder to stop.[34]

It is tempting to think that money already spent on a project would be 'wasted' if you walk away from it. But the rational position on any investment would be to consider only the prospects of it turning a profit against future costs. It should be the same decision regardless of whether you are embarking on a fresh venture or have already invested in one. If a stock in your portfolio is dropping in value, you should sell it, regardless of how much you originally bought it for or how long you've had it. Sure, it may rally and bounce back. But there are also costs to having your capital invested there rather than in some other stock that is likely to continue rising in the near-term future. If humans acted rationally, past investment would not influence future decisions. [35]

If a project is hitting significant problems and the costs are spiralling, the rational response is to cut your losses and walk away. There's no sense in throwing good money after bad. Even if your bloody-mindedness sees the project through to completion, you'll still have made a loss on the venture. And this is exactly what happened with Concorde: the costly development was continued long after it became clear that the economic case had disintegrated and the project was unlikely ever to return a profit.

It's a particularly telling example of a cognitive bias known as the 'sunk cost fallacy': the tendency to continue an endeavour once an investment has been made, even after it becomes apparent that the outcome is unsatisfactory. But it has also become known as the Concorde fallacy.[36] The British and French teams behind the project would have done well to heed the words of comedian W.C. Fields: 'If at first you don't succeed, try, try again. Then quit. No use being a damn fool about it.'

The sunk cost fallacy is also a powerful influence locking nations into ongoing war, long after the possibility of achieving the original objectives has receded and the costs continue to rise. The US involvement in the Vietnam War stretched to twenty years before the fall of Saigon in 1975, spanning the administrations of five presidents (both Republican and Democrat): Dwight D. Eisenhower, John F. Kennedy, Lyndon B. Johnson, Richard Nixon and Gerald Ford. Repeated congressional debates approved the continuation of the war, in part so that the dead would not have died in vain. Yet the longer the war was perpetuated, the more difficult it became to accept the tremendous losses with nothing to show for them.[37] Still, not wanting the sacrifices to have been in vain is particularly perverse reasoning for risking even more lives.

Forty years later, the US once again found itself drawn into a protracted war with no end in sight. Following the September 11th attacks in 2001, the US-led coalition launched its war on terror and invaded Afghanistan, principally to topple the fundamentalist Taliban government and target Osama bin Laden and his al-Qaeda organisation. After sixteen years of military intervention, and still no clear pathway to an ultimate victory, President Trump declared in August 2017 that he would expand the US military presence and keep US boots on the ground indefinitely.[38] While there were good reasons for attempting to secure stability and security in the country, the rhetoric used to justify the continuation of a US presence 'after the extraordinary sacrifice of blood and treasure' very much reflected the sunk cost fallacy: 'our nation must seek an honorable and enduring outcome worthy of the tremendous sacrifices that have been made, especially the sacrifices of lives'.[39] When the United States withdrew its forces in August 2021, the war in Afghanistan had become the longest conflict in US history, surpassing even the Vietnam War. The result was that the Afghan security forces quickly collapsed, and the Taliban reclaimed the country.

LOSS AVERSION

Another deep-seated cognitive bias that influences many of our more complex interactions, and consequently has played a powerful role in international relations and conflicts, is that of loss aversion. There is a fundamental asymmetry in the values we ascribe to equivalent losses and gains. Noticing that you've misplaced (or had stolen) £100 creates a much greater displeasure than the pleasure you experience winning £100 in the lottery. Losses loom larger than gains.[40]

In fact, psychological experiments using economic games have been able to quantify this imbalance. They have found that losses typically weigh 2–2.5 times more heavily on people than gains. As a result, we re-adapt much more quickly to positive changes in our lives than we do to negative ones. For example, our overall happiness returns to a normal level much sooner after a pay rise than a pay cut.

Loss aversion underlies other cognitive biases, such as the 'status quo bias' and the 'endowment effect'.[41] The former manifests itself in our general preference to stick with the current state of affairs rather than pursuing an alternative, because we fear the potential losses of switching more than we anticipate the potential gains; this opposition to change forms the psychological basis of conservatism. The endowment effect describes how we feel more strongly about holding on to an item that we possess than about acquiring such a thing in the first place.

These biases also have a strong influence on the risks we're willing to accept when we have to make a decision in uncertain circumstances. The human appetite for risk varies depending on whether we consider the chances of making a gain or suffering a loss. (There's also variation between personalities, but that's a different issue.) When considering a potential gain, we prefer the option that offers a greater chance of a small gain over the one that offers a less probable but larger reward. We're therefore risk-averse when it comes to making gains. For example, we tend to opt for the chance to win £10 based on a 50/50 coin toss rather than £40 from a one-in-six dice roll. This goes against the strictly economic expectation that a perfectly rational person would pick the dice roll because it has the higher average payout (£6.66 vs. £5).

On the other hand, our preference reverses and we have a tendency to be risk-seeking when we face the prospect of suffering a loss. We pick the option that offers the chance of minimising our loss even if it is less likely to occur compared to the alternative. So if we're already in the red, we'll take bigger gambles in an attempt to recover our losses so far. This tendency towards risky behaviour when we feel like we've been losing compounds the sunk cost fallacy we looked at earlier.

This shift in risk appetite makes sense in an evolutionary context. In the natural environment of our ancestors, losing food or another resource could mean the difference between life and death. When our survival is at stake, it makes sense to take larger risks: desperate times call for desperate measures.[42]

Our understanding of these related cognitive biases – loss aversion, the status quo bias and the endowment effect – has been combined into something of a grand unified theory of how we make decisions in conditions of uncertainty. Prospect theory is founded on experiments of how people actually behave when making decisions and judgements, in contrast to previous models of decision-making based on traditional economic thinking that viewed humans as perfectly rational calculators. It was developed in 1979 out of the long-running research programme on the interface between psychology and economics conducted by Daniel Kahneman and Amos Tversky.[43] Kahneman went on to be awarded the Nobel Prize in Economics in 2002 (Tversky had died six years previously and the Nobel is never awarded posthumously) for his work on human judgement and decision-making under uncertainty that is largely responsible for the creation of the field of behavioural economics.[44]

At the very core of prospect theory lies the human tendency to be loss-averse, and it shows that such cognitive biases have

important consequences, in particular for bargaining and negotiations. Take negotiations of international trade agreements,
for example. These involve discussions between nations to reach
an accord on details relating to the taxes and tariffs that will be
applied on different commodities or manufactured products, as
well as quotas or other restrictions and the penalties for infringe-
ments of the terms. Securing an agreement involves each side
making concessions to the stipulations and demands of the
other. For each nation, the concessions made to the other side
represent losses, whereas the concessions granted by the other
side are gains. But because of loss aversion and the asymmetry
in how we evaluate our own losses and gains, both sides can
feel that their own concessions are not adequately balanced by
those from the other party. The result is that both sides expect
the other to offer up more than they themselves are willing to
concede, and the cognitive bias can often make it difficult for
negotiations to reach a successful conclusion.[45]

The situation is even more acute when a nation's security, or
even survival, is at stake. From the early 1970s, for example, the
superpowers of the US and the USSR (and later Russia) negotiated a series of bilateral agreements on strategic nuclear arms to
reduce the number of warheads, as well as the ballistic missiles
and long-range bombers required for delivering them to targets,
that each had in their arsenal.[46] The talks often hit a stalemate,
and breakthroughs took years to negotiate. Part of the problem
was that their respective weapons systems were not directly
comparable – in terms of their yield, accuracy or range – and
required complicated horse-trading over, for example, how
many of a particular kind of US ballistic missile would fairly
balance those possessed by the USSR. But the psychology of
loss aversion also played a powerful role: each side perceived a
greater loss from the dismantling of their own nuclear missiles
than the gain in security from an equivalent reduction made by
their rival, and so felt disadvantaged by each deal on the table.[47]

The same cognitive biases make it much harder in international relations to pressure another state into discontinuing a

course of action than to enforce their inaction in the first place. The endowment effect means that a nation will resist giving up something it already possesses much more fervently than another nation will seek to acquire it.[48] For example, since the Second World War, eight states have developed nuclear weapons: the US, Russia, the UK, France, China, India, Pakistan and North Korea. Israel is also widely believed to possess a nuclear capability but refuses to acknowledge it, instead pursuing a policy of deliberate ambiguity. So far, the Non-Proliferation Treaty has successfully prevented any of the other nearly 190 sovereign states developing a nuclear weapons programme. However, only one country has developed its own nuclear arsenal and voluntarily given it up – South Africa, which assembled six nuclear weapons under the Apartheid regime but dismantled them all before the election of the African National Congress into government in the early 1990s.[49] (And after the dissolution of the USSR, the former Soviet states of Ukraine, Belarus and Kazakhstan all transferred the nuclear arsenals located in their republics back to Russia.)

Prospect theory emphasises that people tend to evaluate their gains and losses not in absolute terms but relative to a particular reference point, and this is often the status quo – what they currently have. In the case of nuclear disarmament, the US and the USSR at least had a pretty good idea of the numbers of weapons each other possessed, but conflict resolution becomes much more difficult when the two sides have different ideas about the relevant status quo because of a long and convoluted history of possession. This is often the case in territorial disputes which become all the more entrenched and bloody when rivals lay claim to the same land that has changed hands back and forth in the past. In the Middle East, for example, both the Israelis and the Palestinians lay claim to the West Bank on the River Jordan and the Gaza Strip, and each perceives the other as the aggressive intruder on their historical lands. In any attempt at resolution, no matter how the territory is divided, both sides feel an acute loss from their perceived status quo that greatly outweighs gains received from the deal.[50]

A similar situation occurred in Northern Ireland, and here the negotiation of the Good Friday Agreement is a good example of how prospect theory shows such conflicts can be peacefully resolved.

Ireland was partitioned by the British in 1921 to create Northern Ireland with a majority of unionists and loyalists who wished to remain in the United Kingdom and, as descendants of seventeenth-century colonists from Britain, were mostly Protestant. The southern part of the island was designated the Irish Free State – and later became the Republic of Ireland – containing mostly Catholic Irish nationalists who wanted a united, independent Ireland. Tensions persisted, with the significant minority of Catholic nationalists in Northern Ireland feeling discriminated against by the unionist governments. Mounting unrest led to the outbreak of sectarian violence in the late 1960s known as the Troubles. After thirty years of conflict, lasting peace finally seemed possible in April 1998, with a deal between most of Northern Ireland's political parties as well as an arrangement between the British and Irish governments known collectively as the Good Friday Agreement.

The agreement covered a complex set of issues, including those relating to sovereignty, governance, disarmament and the security arrangements in Northern Ireland. The formation of a new assembly provided devolved legislature for Northern Ireland and a devolved executive with ministers from across the political spectrum. Institutions were put into place to ensure greater coordination on policies between the northern and southern jurisdictions, as well as a council of British and Irish ministers to enhance cooperation. And in order to end sectarian violence, paramilitary groups were to commit to the decommissioning of weapons in return for the early release of paramilitary prisoners, reforms to policing practices and a reduction in the presence of the British armed forces to 'levels compatible with a normal peaceful society'.

The Good Friday Agreement was hugely significant as it was mutually acceptable to the two political groups – unionist and nationalist – which had diametrically opposed goals for the end state of Northern Ireland, and it offered a solid chance at peace after one of the longest-running conflicts of modern history. But not only did the opposing political leaders need to reach an accord, they then also had to sell the deal they had brokered to the public for it to be ratified by referendum.[51] So what can prospect theory tell us about why the Good Friday Agreement was a success – in terms of both political accord and popular support?

As we have seen, one major insight of prospect theory is that people tend to be much more strongly motivated by what they could lose in making a decision. So the most effective way to avoid loss aversion sinking an agreement, and indeed to exploit its effects, is to present an option as the best alternative to a probable loss.[52]

The key to the Good Friday Agreement was that both union-ists and nationalists saw it as the best chance for avoiding the loss of the security gained since the paramilitary ceasefire and of significant potential economic improvement. And most import-antly, both the unionist and nationalist leaders were able to frame the agreement in a way that would be supported by their followers. The unionists argued that it was the best solution to avoid yielding sovereignty to the Republic while also strength-ening the union between Northern Ireland and Great Britain. The nationalists, on the other hand, stressed that the agreement delivered equality for all citizens of Northern Ireland, and while not representing the end of the struggle for a united Ireland, it offered better political means for achieving this republican goal. The Good Friday Agreement deliberately didn't engage with the question of the future of Northern Ireland – whether it should remain in the Union or unify with the Republic – so that both sides were able to rally behind it and the discussion could focus on what they stood to lose if the agreement failed.[53]

A month after the agreement had been signed by most of

Northern Ireland's political parties, as well as the British and Irish governments, a referendum was held for it to be publicly ratified. The Yes campaign secured a large majority of the votes in both parts of Ireland, and while the complex implementation of the agreement was not without difficulties, peace has held for over twenty years.

Coda

Through the chapters of this book, we've explored how aspects of our biology – of our innate humanness – have had a defining influence on history.

We saw how our psychological software developed for social living and altruism and how widespread cooperation enabled us to undertake the enormous, coordinated venture that is civilisation. We saw how our unique reproductive behaviour gave rise to the human family, and how dynasties in different cultures addressed the problem of ensuring an heir. We discussed our susceptibility to infections and the ramifications of endemic disease and raging pandemics. We explored the power of demography, the large-scale properties of our populations and the consequences of our proclivity for exploiting psychoactive substances to alter our conscious experiences. We saw specific examples of the historical consequences of defective genes in our DNA. Finally, we looked at the multitude of cognitive glitches and biases that affect our behaviour.

Human history has played out in the balance between our faculties and our flaws as a species.

But we've not been powerless slaves to our biology. Human technological progress is the story of how we have endeavoured to enhance and augment our natural capabilities and compensate for or overcome many of our biological inadequacies.

Lacking the sharp claws or sabre-like teeth of other animals, we used stone tools such as the hand axe or pointed spear to hunt and butcher our prey, enabling us to enrich our diet with meat. The same weapons allowed us to protect ourselves against predators, and we also used them to fight each other.

The exploitation of fire for cooking, in combination with the development of pottery and clay vessels, allowed us to deactivate toxins and preserve and store foodstuffs. It also provided an external pre-digestion system to extract more nutrition from food. Likewise, the quern and then the millstone, grinding and pounding grains into flour, served as a technological extension of our molar teeth.

From the sewing of animal hides into clothing to the innovation of the loom for weaving textiles, we found insulation for our furless bodies and protection from the elements as we left tropical Africa and dispersed around the world to colder climes.

These developments are all aspects of human culture – learned behaviours and practices, passed from individual to individual and down the generations. Indeed, our capacity for cultural evolution is an immensely powerful force that has enabled humanity to surpass many of the constraints of our nature.

As we've explored in depth throughout this book, intrinsic factors of our biology have had a profound effect on the history of our societies and civilisations. But the opposite has also been true: humanity's cultural innovations have left imprints on our genetic make-up. For example, with the domestication of the goat, the sheep and, in particular, the cow in the last 10,000 years, populations across Europe, the Middle East and parts of Africa and Asia began supplementing their diets with milk. Being mammals, we are nourished as infants by our mother's milk, but after weaning we would naturally stop producing lactase, the key enzyme needed to digest milk. But among the modern descendants of those ancient populations that took up dairying, the gene for lactase has adapted to remain switched on through adulthood.[1] We have evolved to be better biologically suited to our own cultural environment. Today, 95 per cent of northern Europeans exhibit 'lactase persistence' – therefore being able to pour milk liberally on their breakfast cereal or into their tea – whereas other populations around the world become sick if they try to drink milk as adults.[2] So, not only have we been using cultural inventions to enhance our biological abilities, but these innovations have in turn changed our biology.

The pace of cultural change has accelerated enormously since the dawn of civilisation. We have developed ever more sophisticated technologies. Metalworking offered tools more versatile than those made of stone, as well as more durable weapons and protective armour for our soft flesh and fragile skulls. The invention of writing enabled us to vastly expand the information we could store beyond the memory capacity of our own brains and our oral traditions, and to communicate ideas across space and time to people we would never meet. Inscriptions on clay tablets, papyrus or parchment gave way to paper and the printing press, ultimately leading to the internet where we can access effectively limitless information from the palm of our hand. We invented spectacles to correct blurred eyesight, as well as the telescope and microscope to extend our vision into otherwise invisible realms. Modern medicine of antibiotics, vaccines and prophylactics support our immune system and protect us from disease; other pharmaceuticals mask the effects of genetic defects, while skilful surgery fixes the consequences of anatomical deformities or injuries. We've also been able to take control of our own reproduction: the use of condoms, hormonal birth-control pills and other contraceptives, as well as medically safe abortions, decouples sex from procreation and provides the choice over when and with whom we want to have children, thus allowing us to regulate family size and population growth. Modern technology can also aid reproduction when we encounter difficulties; in vitro fertilisation, for example, offers hope to otherwise infertile couples.

Through all these innovations, and many more, we've come to possess the power to complement our natural abilities and compensate for our limitations. So much so that today, in the developed world at least, the differential survival rate or reproductive success of individuals is rarely dependent on their genetics. Natural selection no longer has the raw material to operate, and the evolution of our species has effectively ceased.[3]

Most of us now live in an environment almost entirely created and controlled by ourselves. That is not to say, however, that we are no longer subject to the prescriptions of our biology.

Let's look at just a couple of examples. In our modern urban society, fewer and fewer people work on fields and in factories. For many of us, whether we work in finance or in call centres, we sit down for long hours of the day, hunched over a desk. After countless generations of roaming hunter-gatherers, agricultural labourers and industrial workers, we are now almost entirely sedentary.

Indeed, we now spend so much of our working day sitting down – or even worse, so much of our time off slouched on the sofa – that the key postural muscles engaged in supporting our spine and holding us upright when we stand are atrophied. The vast majority of us will suffer from chronic back pain, specifically in the lower back, at some point in our lives. Moreover, near universal access to cars and public transport in the developed world strips us of the need for physical exertion in our daily comings and goings.

Foragers and farmers spend their lives active on their feet, but with the rise of the modern industrialised world an odd concept emerged – taking exercise. As day-to-day life has become so bereft of exertion, we have purposefully inserted it back into our routine. For the ancient Greeks in their gymnasiums (derived from *gymnos*, the Greek word for 'naked'), exercising was a pastime for the privileged social classes with slaves who didn't need to labour. Today, we rush to the gym before or after work to stay fit. We can only guess what medieval peasants would have made of the apparent absurdity of paying to walk or run on the spot, pounding out virtual miles on the treadmill (used as a form of punishment by hard labour in early-nineteenth-century British prisons, incidentally).

The shift to a largely indoor existence has also affected our vision. Eyesight tends to deteriorate with age as the lens of the eye becomes less flexible, making it harder to focus on nearby objects, meaning people tend to become long-sighted as they get older. But since the Victorian age there has been a phenomenal surge of the opposite problem – myopia, or short-sightedness – even among children. Our lives have become so dominated by looking close-up – especially at screens – and rarely spending time outside, scanning the distance, that myopia now affects up to 50 per cent of those living in urban environments.[4]

Modern life in the developed world has also led to a proliferation of allergic reactions, such as asthma, eczema, food allergies and hay fever (which is perhaps ironic considering that we evolved as a grassland species). These conditions are caused by the inflammation of soft tissues in our body and represent the immune system overreacting to innocuous triggers. While hygiene is key for reducing transmissible diseases, we have become overly cautious in keeping our homes fastidiously clean and stopping infants playing outside in the dirt. Yet the immune system needs to train itself to distinguish between genuine threats and harmless stimuli, and without early exposure to dust, bacteria and parasites, it doesn't learn properly and so becomes hypersensitive and prone to allergic reactions.

What's more, the leading causes of death in the developed world are no longer pestilence or famine but largely preventable conditions that we bring upon ourselves: obesity, diabetes, high blood pressure, heart disease. The problem behind these 'lifestyle diseases' is partly our effortless, immobile existence, but also the gluttony of fabulously rich foods we eat. We are facing epidemics not of infectious diseases but of the avoidable consequences of overconsumption and physical inactivity. Industrialised agriculture, with enormously efficient mechanisation and artificial fertilisers, pesticides and herbicides, provides an abundant harvest, and the mass-production of meat means it has also never been so cheap or more widely available. We live in a time of overwhelming bounty. But the issue is not just the amount of food at our disposal but the kind of meals we often choose to eat. On the whole, we don't tend to consume unhealthy amounts of fresh fruit and veg. And the root cause of our overindulgence comes down to programming deep in our biological make-up.

For our ancestors in the African savannah, survival required carefully directed effort, and evolution therefore programmed our sense of taste to favour sources of vital nutrients and minerals that were scarce in this environment, such as sugar, fat and salt. But because human evolution cannot now keep up with cultural change, we have kept our palaeolithic palate and still crave the high-value foods which are today available in abundance.

Seen in this light, the totem of modern fast food – a cheeseburger with fries and a soft drink – is like an ancestral dream come true: greasy protein, encased in energy-rich carbs, sprinkled with moreish salt and washed down with concentrated sugar solution. It is almost uncannily well-crafted to hit each and every one of our primal dietary drives and light up the pleasure centres of our brain. These food types activate the dopamine reward pathway in the brain in just the same way as the other addictive substances we explored earlier.[5] Eating such rich food feels not only satisfying in the short term, but compulsively so. Indeed, almost all the modern processed food we consume is loaded with fat, salt and sugar. And much of the meat we eat has been industrially ground into mince – we've even outsourced the effort of chewing to a factory.

In terms of providing calories for the human body, the energy-packed, soft, easily digestible nutrients of modern meals are effectively rocket fuel supplied to people who are barely mobile. There is a sharp disconnect between the ancestral environment in which our genes evolved and the modern world we have created for ourselves. So when we allow our ancestral urges to determine our diet, we become prone to so-called mismatch diseases. Surplus energy is stored by the body in fat reserves, causing obesity; excessive salt drives high blood pressure, contributing to heart disease; and the spikes in blood sugar levels result in diabetes.

Another major reason why we find it so hard to resist eating too much processed food and sweet treats, despite knowing full well they are unhealthy and fattening, comes down to a cognitive bias. We fail to act rationally by overvaluing immediate rewards while disregarding the longer-term consequences of our choices. This tendency makes evolutionary sense. In an uncertain or dangerous world, it pays to seize benefits immediately as you may not get the chance later, or to focus on a current threat rather than something that lies over the horizon. And in the modern world, the 'present bias' – a cognitive short-sightedness – rears its head not only in unwholesome eating habits, but also in opting to spend spare cash today rather than building up savings for the future, or in procrastinating by seeking instant gratification now and putting off chores or work until later (if ever!). It is also one of several cognitive biases that hinders us in responding effectively to serious but gradually developing problems such as climate change.

The fact that the Earth's climate is warming due to human activity is scientifically very well established, and if we don't act quickly and decisively, the consequences could be extremely dire indeed. Effective solutions to the problem require not only each of us to alter our individual behaviour to reduce greenhouse gas emissions, but also governments and industries to instigate top-down changes to policy and practices (themselves responding to what we as voters and consumers indicate we want). And I think it's fair to say that most of us are now well aware of the changes we should be making to our current lifestyles. The problem is, we need to sacrifice immediate benefits – such as driving a big, comfortable car, flying away on summer holidays or enjoying meat or dairy products – for the sake of preserving the environment for the long-term future.

(Although, perhaps most alarming about climate change is just how apparent the effects have already become in the past few years.) Even if we will personally gain from lifetime savings on fuel bills, the present bias deters us from buying energy-efficient appliances because of the higher upfront cost.

The sunk cost fallacy, which we explored in Chapter 8, also plays into the problem. The more time, energy or resources we've already committed to a particular approach, the more likely we are to stick with it even after it has become clear it would be beneficial to change tack. This cognitive bias is part of the reason our infrastructure continues its reliance on fossil fuels despite mounting evidence of the advantages of renewable or carbon-neutral alternatives.

And this is assuming, of course, that you accept the severity of the issue: a sizeable fraction of the public still does not believe in climate change. Most of us get our news from mass media outlets, which have become increasingly polarised in their ideological stances. Confirmation bias only serves to strengthen the conviction of those disbelieving the severity of the situation.* We have seemingly been hardwired with a number of cognitive biases that impede our ability to take appropriate action to address seemingly distant, gradual and complex challenges such as climate change.

Cognitive biases, along with many other aspects of our biology and evolutionary past, have had a huge influence on the course of human history. And they still hold a powerful influence on the future we will create.[7]

---

* In one study, for example, people who accepted or rejected climate change were asked to read two articles, one presenting the prevailing scientific consensus and the other a sceptical viewpoint. The acceptors perceived the first article as more reliable, whereas the rejectors found the reverse; but crucially, both groups came away more convinced of their pre-existing views.[6]

List of Figures

Page 252

The Charge of the Light Brigade figure created by the author. Terrain map taken from *The Destruction of Lord Raglan* by Christopher Hibbert (Longman, 1961), available from Stephen Luscombe at http://www.britishempire.co.uk. Troop positions taken from this reference, as well as p.451 of *Our Fighting Services* by Evelyn Wood (Cassell, 1916).

Endnotes

Page numbers given for books; for ebooks, the location is indicated as 'loc.'

INTRODUCTION

1. Collins (2006); White (2020). • 2. National Safety Council (2022). • 3. Lents (2018), loc.340. • 4. Darwin (1859), ch.6. • 5. Yu (2016), p.31. • 6. Steele (2002). • 7. Marcus (2008), p.107. • 8. 'Phoneme', in Brown, K. (ed.). *Encyclopedia of Language & Linguistics* (Second Edition). Elsevier. • 9. Maddieson (1984). • 10. Pereira (2020).

CHAPTER 1

1. Several recent books provide a good introduction to these human adaptations that enable us to live in peaceable societies, and formed the basis of this chapter: Sapolsky (2017); Christakis (2019); Wrangham (2019); Raihani (2021). • 2. Wrangham (2019), p.180. • 3. Mitani (2010); Wilson (2014). • 4. Wrangham (2019), loc.350. • 5. Wrangham (2019), loc.2804, loc.2825. • 6. Johnson (2015). • 7. Wrangham (2019), loc.2120. • 8. Raihani (2021), p.226. • 9. Christakis (2019), loc.5860. • 10. Wrangham (2019), loc.640; Kruska (2014). • 11. Wrangham (2019), loc.1112, loc.1410. • 12. Wrangham (2019), loc.1415; Theofanopoulou (2017). • 13. Wrangham (1999). • 14. Spiller (1988); Glenn (2000); Jones (2006); Strachan (2006); Engen (2011). • 15. Singh (2022). • 16. Powers

(2014); Mattison (2016). • 17. Anter (2019). • 18. Stewart-Williams (2018), loc.749. • 19. Dugatkin (2007). • 20. Stewart-Williams (2018), loc.4258; Cartwright (2000); Burton-Chellew (2015). • 21. Visceglia (2002); Vidmar (2005). • 22. Trivers (1971); Trivers (2006); Schino (2010). • 23. Raihani (2021), p.133. • 24. Stewart-Williams (2018), loc.4613; de Waal (1997); Jaeggi (2013); Dolivo (2015); Voelkl (2015). • 25. Raihani (2021) p.134. • 26. Raihani (2021) p.134. • 27. Stewart-Williams (2018), loc.620; Massen (2015). • 28. Raihani (2021), p137. • 29. Cosmides (1994); Christakis (2019), loc.4780. • 30. Christakis (2019) loc.5168; Winston (2003), p.313; Stewart-Williams (2018), loc.4780. • 31. Alexander (2020); Nowak (2006); Nowak (2005). • 32. Sapolsky (2017), p.633. • 33. Haidt (2007). • 34. Wrangham (2019), loc.3702. • 35. Edwardes (2019), p.112; Jensen (2007). • 36. Christakis (2019), loc.5209; Fehr (2002). • 37. Wolf (2012). • 38. Kurzban (2015); Raihani (2021), p.163. • 39. Kurzban (2015); Yamagishi (1986); Fehr (2002). • 40. Sapolsky (2017), p.610; de Quervain (2004) • 41. Kahneman (2012), p.308. • 42. Kahneman (2012), p.308. • 43. Sapolsky (2017), p.636. • 44. Edwardes (2019). • 45. Christakis (2019), loc.5280; Fehr (2002); Boyd (2003); Fowler (2005); Boyd (2010). • 46. Dunbar (1992). • 47. McCarty (2001). • 48. But see Linderfors (2021) for a discussion on how trying to place any specific number on this limit is problematic. • 49. Zhou (2005). • 50. Carron (2016). • 51. Dunbar (2015). • 52. Fuchs (2014). • 53. Wason (1968); Wason (1983). • 54. Winston (2003), p.334; Wason (1983). • 55. Winston (2003), p.334. • 56. Cosmides (1989) Cosmides (2010); Haselton (2015). • 57. Cosmides (2015). • 58. Cosmides (1989). Cosmides (2010). • 59. Atran (2001); Stone (2002); Carlisle (2002); Pietraszewski (2021). • 60. Wrangham (2019), loc.3718. • 61. Haidt (2007). • 62. Krebs (2015). • 63. Krebs (2015); Christakis (2019), loc.4066. • 64. Raihani (2021), p.118. • 65. Kanakogi (2022). • 66. Fernández-Armesto (2019), loc.1950; Roth (1997). • 67. Jones (2015). • 68. Raihani (2021), p163; Greif (1989). • 69. Luca (2016); Holtz (2020); Chamorro-Premuzic (2015); Raihani (2021), p.163. • 70. Morris (2014).

CHAPTER 2

1. Gruss (2015); Trevathan (2015). • 2. van Leengoed (1987). • 3. Kendrick (2005). • 4. Lee (2009). • 5. Acevedo (2014); Fisher

(2006). • 6. Schmitt (2015); Flinn (2015); Young (2004). • 7. Christakis (2019), p.179. • 8. Hanlon (2020). • 9. Schmitt (2015); Fisher (1989). • 10. Raihani (2021), p.47. • 11. Campbell (2015). • 12. Hareven (1991). • 13. 'heirloom, n.'. OED Online. December 2022. Oxford University Press. https://www.oed.com/view/Entry/85516 • 14. Duindam (2019), loc.1411. • 15. Rady (2017), loc.464. • 16. Kenneally (2014), p.192. • 17. Shammas (1987). • 18. 'Patriarchy' in Ritzer (2011). • 19. Archarya (2019); Duindam (2019), loc.3010; Hartung (2010); Fortunato (2012). • 20. Wilson (1989); Price (2014). • 21. Barboza Retana (2002). • 22. Haskins (1941); Brewer (1997). • 23. Hrdy (1993). • 24. Herre (2013); Economist Intelligence (2022). • 25. Duindam (2015), loc.340. • 26. Bartlett (2020). • 27. Rady (2020), loc.208. • 28. Rady (2017), loc.477. • 29. Rady (2017), loc.510; Rady (2020), loc.1105. • 30. Bartlett (2020), loc.4085–4261. • 31. Rady (2017), loc. 510. • 32. Rady (2017), loc.,430; Rady (2017), loc.750. • 33. Rady (2017), loc.530; Rady (2020), loc.1340. • 34. Rady (2020), loc.1533. • 35. Rady (2017), loc.530. • 36. Rady (2020), loc.1120. • 37. Bartlett (2020). • 38. Rady (2017), loc.425. • 39. Rady (2020), loc.380. • 40. Rady (2020), loc.377; Rady (2020), loc.1340. • 41. Rady (2017), loc.530; Rady (2020), loc.110. • 42. Rady (2017), loc.840. • 43. Murdock (1962); White (1988); Stewart-Williams (2018), loc.3860; Schmitt (2015); • 44. Schmitt (2015) • 45. Campbell (2015). • 46. Duindam (2019), loc.675. • 47. Christakis (2019), loc.2740; Duindam (2019), loc.670; Starkweather (2012); Schmitt (2015). • 48. Christakis (2019), loc.2750; Monaghan (2000), loc.1290. • 49. Christakis (2019), loc.2585. • 50. Zimmer (2019), loc.3210. • 51. Meekers (1995). • 52. Payne (2016); Scheidel (2009a). • 53. Duindam (2019), loc.670. • 54. Schmitt (2015); Stewart-Williams (2018), loc.5550. • 55. Christakis (2019), loc.2390; MacDonald (1995); Scheidel (2009a). • 56. Payne (2016); Betzig (2014); Scheidel (2009a). • 57. Christakis (2019), loc.2395; Payne (2016). • 58. Christakis (2019), loc.2396. • 59. Stewart-Williams (2018), loc.3870. • 60. Kramer (2020). • 61. Archarya (2019). • 62. Duindam (2019), loc.3497. • 63. Kokkonen (2017). • 64. Montesquieu (1777). • 65. Kokkonen (2017); Duindam (2019), loc.3590. • 66. Kokkonen (2017). • 67. Peirce (1993), p.46. • 68. Payne (2016). • 69. Duindam (2019), loc.3235. • 70. Duindam (2019), loc.3300, loc.3310. • 71. Betzig (2014). • 72. Bartlett

(2020). • 73. Duindam (2019), loc.3250. • 74. Betzig (2014); Xue (2005). • 75. Zerjal (2003); Betzig (2014). • 76. Bartlett (2020). • 77. Duindam (2019), loc.3680. • 78. Duindam (2019), loc.2830. • 79. Duindam (2019), loc.3760, loc.788. • 80. Duindam (2019), loc.3295. • 81. Duindam (2019), loc.3718. • 82. Peirce (1993), p.46. • 83. Betzig (2014). • 84. Dale (2017); Dale (2018), p.2. • 85. Betzig (2014). • 86. Duindam (2019), loc.6055. • 87. Betzig (2014). • 88. Duindam (2019), loc.3450; Betzig (2014). • 89. Bixler (1982); Scheidel (2009). • 90. Christakis (2019), loc.3445; Hegalson (2008). • 91. Rady (2020), loc.1662; Alvarez (2009); Helgason (2008). • 92. Vilas (2019). • 93. Rady (2020), loc.1670. • 94. Vilas (2019). • 95. Vilas (2019). • 96. Rady (2020), loc.1670; Zimmer (2019), loc.310. • 97. Rady (2020), loc.1670. • 98. Alvarez (2009). • 99. Alvarez (2009). • 100. Alvarez (2009). • 101. Zimmer (2019), loc.410; Stanhope (1840), p.99. • 102. Alvarez (2009). • 103. Rady (2020), loc.1919. • 104. Rady (2020), loc.2920. • 105. Rady (2020), loc.2918. • 106. Zimmer (2019), loc.420. • 107. Falkner (2021). • 108. Bartlett (2020). • 109. Duindam (2019), loc.1984. • 110. *The Boston Globe* (2021). • 111. Duindam (2019), loc.2005. • 112. Hess (2015); *The Boston Globe* (2021). • 113. Landes (2004); Duindam (2019), loc.2066.

CHAPTER 3

1. Badiaga (2012); Holmes (2013). • 2. Schudellari (2021); Khateeb (2021). • 3. Lacey (2016). • 4. Taylor (2001). • 5. Taylor (2001). • 6. Gurven (2007). • 7. Martin (2015), loc.208. • 8. Sharp (2020); Monot (2005). • 9. Crawford (2009), ch5. • 10. Clark (2010), p.50; Webber (2015), loc.2976; Martin (2015), loc.1770. • 11. Carroll (2007). • 12. Winegard (2019), loc.2334. • 13. Webb (2017). • 14. Phillips-Krawczak (2014); Depetris-Chauvin (2013); McNeill (1976); Yalcindag (2011). • 15. Winegard (2019), loc.2774. • 16. Gianchecchi (2022). • 17. Winegard (2019), loc.370, loc.2774. • 18. Crawford (2009), ch.5. • 19. Acemoglu (2001); Bryant (2007); Gould (2003). • 20. Winegard (2019), loc.3628. • 21. Winegard (2019), loc.3640. • 22. Winegard (2019), loc.3600. • 23. Green (2017); Martin (2015), loc.2780. • 24. Green (2017). • 25. Whatley (2001) • 26. Miller (2016); Winegard (2019), ch.10; Carroll (2007); Armitage (1994);

McNeill (2015), pp.105–123. • 27. Winegard (2019), loc.4190; McNeill (2010), p.201. • 28. Winegard (2019), loc.2847. • 29. Guerra (1977). • 30. Achan (2011); Foley (1997). • 31. Winegard (2019), loc.4184. • 32. Sherman (2005), p.347. • 33. Winegard (2019), loc.4340. • 34. Winegard (2019), loc.4345. • 35. Sherman (2005), p.348. • 36. Sherman (2005), p.349. • 37. Winegard (2019), loc.4400; McNeill (2010), p.222. • 38. Sherman (2005), p.349; Winegard (2019), loc.4240; McNeill (2010), p.199. • 39. Winegard (2019), loc.4250; McNeill (2010), p.199; McCandless (2007). • 40. McNeil (2010), ch.7; Winegard (2019), ch.13. • 41. Watts (1999), p.235. • 42. Watts (1999), p.235. • 43. Winegard (2019), loc.4530. • 44. Oldstone (2009), loc.172. • 45. Sherman (2005), p.341. • 46. Depetris-Chauvin (2013); Webb (2017); Winegard (2019), loc.607; Webber (2015), loc.763; Doolan (2009). • 47. Nietzsche (1888). • 48. Mohandas (2012). • 49. Webb (2017); He (2008). • 50. Meletis (2004); Parsons (1996). • 51. Weatherall (2008). • 52. Mitchell (2018); Randy (2010). • 53. Dyson (2006); Lichtsinn (2021); Dove (2021). • 54. Williams (2011); Gong (2013). • 55. Webber (2015), loc.795; Akinyanju (1989); Dapa (2002). • 56. Malaney (2004). • 57. Kato (2018). • 58. Pittman (2016); Swerdlow (1994); Glass (1985). • 59. Pittman (2016); Webber (2015), loc.2324; Galvani (2005); Dean (1996); Stephens (1998); Lalani (1999); Novembre (2005). • 60. Josefson (1998); Poolman (2006). • 61. Sherman (2005), p.341. • 62. Sherman (2005), p.341; Zinsser (1935), p.160; Clark (2010), p.237; Winegard (2019), loc.4637. • 63. Winegard (2019), ch.13; McNeill (2010), ch.7. • 64. Girard (2011). • 65. Winegard (2019), loc.4656. • 66. Oldstone (2009), loc.175; Winegard (2019), loc.4660. • 67. Winegard (2019), loc.4606; Sherman (2007), p.147. • 68. Modern equivalent calculated using the CPI Inflation Calculator, available at: www.officialdata.org • 69. Bush (2013). • 70. This section on the long-term consequences of the disease environment and extractive and settler colonies is based on Acemoglu (2001). • 71. Clark (2010), p.122. • 72. Acemoglu (2001). • 73. Bernstein (2009), loc.4747. • 74. Esposito (2015); Bernstein (2009), loc.4755 • 75. Winegard (2019), loc.2852. • 76. Winegard (2019), loc.2861. • 77. Morris (2014), p.220. • 78. Morris (2014), p.195. • 79. Winegard (2019), loc.2658. • 80. Winegard (2019), loc.2848. • 81. Roberts (2013), p.792; Sherman (2007), p.324. • 82. Sherman (2007), p.322 • 83. Roberts (2013), p794.

CHAPTER 4

1. Diamond (1987). • 2. Martin (2015), loc.257; Oldstone (2009), loc.2176. • 3. Webber (2015), loc.2266. • 4. Martin (2015), loc.257; Stone (2009). • 5. Grange (2021). • 6. Clark (2010), p.115; Smith (2003) • 7. Clark (2010), p.125. • 8. Clark (2010), p.115. • 9. Outram (2001). • 10. Outram (2001); Parker (2008). • 11. Harrison (2013). • 12. Green (2017). • 13. Crawford (2009), ch.3. • 14. Thucydides. *The History of the Peloponnesian War*, Book II, Chapter VII. Translated by Richard Crawley (1874). Available from Project Gutenberg: https://www.gutenberg.org/files/7142/7142-h/7142-h.htm • 15. Martin (2015), loc.540. • 16. Martin (2015), loc.690; Crawford (2009), ch.3. • 17. Martin (2015), loc.755. • 18. Crawford (2009), ch.3. • 19. Martin (2015), loc.760; Crawford (2009), ch.3. • 20. Crawford (2009), ch.3; Martin (2015), loc.770; Alfani (2017). • 21. Martin (2015), loc.770; Alfani (2017). • 22. Harper (2017), ch.4. • 23. Winegard (2019) loc.1570; Harper (2015); Huebner (2021). • 24. Harper (2015). • 25. Clark (2010), p.166. • 26. Harper (2015). • 27. Sherman (2005), p.60. • 28. Harper (2015). • 29. Crawford (2009), ch.3. • 30. Martin (2015), loc.832; Alfani (2017). • 31. Harbeck (2013). • 32. Green (2017). • 33. Clark (2010), p.91. • 34. Martin (2015), loc.880. • 35. Winegard (2019), loc.1680. • 36. Winegard (2019), loc.1680; Crawford (2009), ch.3. • 37. Alfani (2017); Webber (2015), loc.1302. • 38. Alfani (2017). • 39. Martin (2015), loc. 965; Alfani (2017). • 40. Eisenberg (2019). • 41. Alfani (2017). • 42. Martin (2015), loc.930; Sarris (2007). • 43. Alfani (2017); Sarris (2002); Sarris (2007). • 44. Martin (2005), loc.938. • 45. Alfani (2017). • 46. Alfani (2017). • 47. Eisenberg (2019). • 48. Martin (2005), loc.940; McNeill (1976), p.123 • 49. Martin (2005), loc.940. • 50. Shahraki (2016). • 51. Clark (2010), p.91. • 52. Sarris (2002); Mitchell (2006); Harper (2017); Little (2006). • 53. Green (2017). • 54. Wheelis (2002). • 55. Webber (2015), loc.1240; Martin (2002), loc.855. • 56. Martin (2002), loc.850. • 57. Martin (2002), loc.850; 'bubo, n.'. OED Online. Oxford University Press. https://www.oed.com/view/Entry/24087. • 58. Martin (2002), loc.840. • 59. Alfani (2017). • 60. Martin (2002), loc.1240. • 61. Webber (2015), loc.1260 • 62. Martin (2002), loc.855. • 63. Sussman (2011). • 64. Webber (2015), loc.1270; Alfani (2017). • 65. Clark (2010),

p.218. • 66. Alfani (2017); Pamuk (2007); Pamuk (2014). • 67. Alfani (2017). • 68. Clark (2010), p.218. • 69. Martin (2002), loc.1486. • 70. Clark (2010), p.221. • 71. Alfani (2017); Clark (2010), p.221. • 72. Alfani (2017); Clark (2010), p.221. • 73. North (1970). • 74. Herlihy (1997), p.39. • 75. Herlihy (1997), p.48. • 76. Alfani (2013). • 77. Alfani (2017). • 78. Alfani (2017); Alfani (2017). • 79. Webber (2015), loc.1287; Brook (2013), p.254. • 80. Crawford (2009), ch.5; Sherman (2007), p.53. • 81. Winegard (2019), loc.2518; Crawford (2009), ch.5. • 82. Sherman (2007), p.53. • 83. Oldstone (2009), loc.705; Crawford (2009), ch5; Clark (2010), p.200. • 84. Watts (1999) p.90; Clark (2010), p.200. • 85. Crawford (2009), ch.5. • 86. Webber (2015), loc.1658. • 87. Oldstone (2009), loc.700. • 88. Crawford (2009), ch.5. • 89. Winegard (2019), loc.2505. • 90. Green (2017); Hopkins (2002). • 91. Loades (2003); Webber (2015), loc.1342. • 92. Sherman (2005), p.198. • 93. Sherman (2005), p.198; Crawford (2009), ch.4; Webber (2015), loc.1351. • 94. Oldstone (2009), loc.146; Hopkins (2002). • 95. Webber (2015), loc.1359; Oldstone (2009), loc.146, loc.734; Sherman (2005), p.198. • 96. Ellner (1998). • 97. Clark (2010), p.200. • 98. Koch (2019); McNeill (1976). • 99. McNeill (1976); Yalcindag (2012). • 100. Koch (2019). • 101. Martin (2002), loc.1568. • 102. Crawford (2009), ch.5. • 103. Wallace (2003). • 104. Kuitems (2022). • 105. Mühlemann (2020). • 106. Koch (2019); McEvedy (1977). • 107. Dobyns (1966); Koch (2019). • 108. Nunn (2010); Koch (2019); Ord (2021), p.124. • 109. Koch (2019); Denevan (1992); Denevan (2010); Alfani (2013). • 110. Ord (2021), p.124. • 111. Crawford (2009), ch.5; Martin (2015), loc.2705; Winegard (2019), loc.2541. • 112. Darwin (1839), ch.12. • 113. Webber (2015), loc.2459. • 114. Crawford (2009), ch.5. • 115. Sherman (2007), p86; Majander (2020). • 116. Gobel (2008); Vachula (2019). • 117. Elias (1996); Jakobsson (2017); Marks (2012). • 118. Levy (2009), p.106; Coe (2008), p.193. • 119. Walter (2017). • 120. Martin (2015); Martin (2002); Mackowiak (2005). • 121. McNeill (1976); Diamond (1998) • 122. Phillips-Krawczak (2014). • 123. Crawford (2009), ch.5. • 124. Winegard (2019), loc.2751. • 125. Winegard (2019), loc.2944. • 126. Spinney (2017), p.2. • 127. Martin (2015), loc.2882; Dobson (2007), p.176. • 128. Martin (2015), loc.2882. • 129. Clark (2010), p.243; Honigsbaum (2020), ch.1. • 130. Spinney (2017), ch.3; Honigsbaum

(2020), ch.1. • 131. Ewald (1991). • 132. Taubenberger (2006). • 133. Webber (2015), loc.1946. • 134. Oxford (2018); Spinney (2017), ch.14; Honigsbaum (2020), ch.1. • 135. Spinney (2017), ch.14. • 136. Ewald (1991). • 137. Spinney (2017), ch.12. • 138. Taubenberger (2006). • 139. Oldstone (2009), loc.4743. • 140. Oldstone (2009), loc.4745. • 141. Spinney (2017), p.2. • 142. Spinney (2017), p.2. • 143. Oldstone (2009), loc.4682. • 144. Ayres (1919), p.104 (Diagram 45). • 145. Oldstone (2009), loc.4684. • 146. Oldstone (2009), loc.4690. • 147. Oldstone (2009), loc.4690; Spinney (2017), ch.20. • 148. Stevenson (2011), p.91. • 149. Oldstone (2009), loc.4690; Kolata (2001), p.11. • 150. Zabecki (2001), pp.237, 275. • 151. Stevenson (2011), p.91; Watson (2014), p.528. • 152. Watson (2015), p.339. • 153. Watson (2015), p.528. • 154. Oldstone (2009), loc.2995. • 155. Noymer (2009); Chandra (2012). • 156. Chandra (2014) • 157. Nambi (2020). • 158. Chunn (2015). • 159. Spinney (2017), ch.20; Kapoor (2020). • 160. Chunn (2015), p.207. • 161. Chunn (2015), p.189. • 162. Spinney (2017), ch.20; Arnold (2019). • 163. Chunn (2015), p.190. • 164. Spinney (2017), ch.20; Chunn (2015); Kapoor (2020).

CHAPTER 5

1. Robson (2006). • 2. Galdikas (1990). • 3. Kramer (2019); Lovejoy (1981). • 4. Kramer (2019); Gurven (2007); Hill (2001). • 5. Kramer (2019). • 6. Bowles (2011). • 7. Marklein (2019). • 8. Armelagos (1991). • 9. Diamond (2003). • 10. Zahid (2016); Bettinger (2016). • 11. Li (2014); Bostoen (2018); Bostoen (2020). • 12. Diamond (2003). • 13. Bostoen (2020). • 14. de Filippo (2012). • 15. de Luna (2018). • 16. Reich (2018), loc.3622; Rowold (2016); Bostoen (2020); Holden (2002). • 17. Reich (2018), loc.3622. • 18. Bostoen (2018). • 19. de Filippo (2012); Rowold (2016). • 20. Reich (2018), loc.3622; Bostoen (2018). • 21. Bostoen (2018). • 22. Bostoen (2020). 23. Ehret (2016), p.113. • 24. Webb (2017); Dounias (2001); Yasuoka (2013). • 25. Bostoen (2018). • 26. Tishkoff (2009). • 27. An extensive review of the genetic evidence is provided in Pakendorf (2011). • 28. Bostoen (2020). • 29. de Luna (2018); Bostoen (2018); Bernie ll-Lee (2009). • 30. Bostoen (2018). • 31. de Luna (2018) provides a concise overview of the

history of research into the Bantu Expansion, complete with further references on developments within linguistic, archaeological and genetic studies. For the linguistic analysis of the Bantu Expansion see Cavalli-Sforza (1994) and Rowold (2016). • 32. Gartzke (2011). • 33. Beare (1964). • 34. Knowles (2005). • 35. von Clausewitz, C. (1832) *On War*, Book III, Chapter VIII. • 36. Morland (2019), loc.308. • 37. Zamoyski (2019); Roberts (2015); Tharoor (2021). • 38. Gates (2003), p.272. • 39. Clodfelter (2008). • 40. Morland (2019), loc.799; Office of Population Research (1946) • 41. Blanc (2021). • 42. Beckert (2007). • 43. Desan (1997). • 44. Desan (1997). • 45. Grigg (1980), p.52. • 46. Cummins (2009). • 47. Morland (2019), loc.1284. • 48. Wrigley (1985). • 49. Data on historical birth rates provided by www.statista.com, eg: https://www.statista.com/statistics/1037303/crude-birth-rate-france-1800-2020/ • 50. Clark and Alter (2010). • 51. Cummins (2009). • 52. Morland (2019), loc.820. • 53. Morland (2019), loc.917. • 54. Perrin (2022). • 55. Beckert (2007). • 56. Historical population data for France, UK and Germany provided by www.ourworldindata.org • 57. Morland (2019), loc.1510. • 58. Information from Census 2021 in Northern Ireland available from Northern Ireland Statistics and Research Agency at https://www.nisra.gov.uk/statistics/census/2021-census; Compton (1976); Anderson (1998); Gordon (2018); Carroll (2022); Morland (2019). • 59. BBC News (2022). • 60. Brainerd (2016); Glantz (2005). • 61. Glantz (2005), p.546. • 62. Brainerd (2016). • 63. Brainerd (2016). • 64. Brainerd (2016); Ellman (1994). • 65. Vishnevsky (2018). • 66. Vishnevsky (2018). • 67. Brainerd (2016); Ellman (1994). • 68. Strassman (1984). • 69. Data on sex ratios in India and China provided by www.statista.com • 70. Central Intelligence Agency (2021). • 71. Brainerd (2016). • 72. Brainerd (2016); Sobolevskaya (2013). • 73. Bethmann (2012). • 74. Pedersen (1991); Schacht (2015). • 75. Kesternich (2020). • 76. Bethmann (2012). • 77. Gao (2015). • 78. Fernandez (2004). • 79. Bethmann (2012). • 80. Manning (1990), p.104; Teso (2019); Nunn (2017) • 81. Manning (1990), p.85; Nunn (2017). • 82. Nunn (2008). • 83. Nunn (2010). • 84. Nunn (2011); Nunn (2017); • 85. Nunn (2008); Green (2013); Whatley (2011). • 86. Zhang (2021). • 87. Teso (2019); Lovejoy (2000); Lovejoy (1989) • 88. Teso (2019); Lovejoy (1989). • 89. Teso (2019); Thornton (1983); Manning (1990). • 90. Teso (2019). • 91.

Teso (2019). • 92. Teso (2019). • 93. Manning (1990); Edlund (2011); Dalton (2014); Bertocchi (2015). • 94. Bertocchi (2019) • 95. Bertocchi (2019). • 96. Ciment (2007); Nunn (2008); Nunn (2010); Zhang (2021). • 97. Winegard (2019), loc.4425. • 98. Winegard (2019), loc.4426. • 99. Simpson (2012), ch.1. • 100. Hill (2008), p.90, p.140. • 101. Simpson (2012), ch.2. • 102. Godfrey (2018). • 103. Grosjean (2019). • 104. Grosjean (2019). • 105. Grosjean (2019). • 106. Pedersen (1991). • 107. Behrendt (2010). • 108. Grosjean (2019); Raihani (2021), p.58.

CHAPTER 6

1. Campbell-Platt (1994). • 2. Jennings (2005). • 3. McGovern (2018). • 4. Hames (2014), p.6. • 5. Katz (1986). • 6. Dominy (2015). • 7. Standage (2006), p.23. • 8. Hames (2014), p.10. • 9. Phillips (2014), p.4. • 10. Philips (2014), p.4. • 11. Doig (2022), p.257; Carrigan (2014). • 12. Brooks (2009). • 13. Doig (2022), p.257; Edenberg (2018); Hurley (2012). • 14. Doig (2022), p.260. • 15. Miron (1991). • 16. Bostwick (2015). • 17. Toner (2021). • 18. Macdonald (2004). • 19. Sapolsky (2017), p.64. • 20. Bowman (2015). • 21. Sapolsky (2017), p.65. • 22. Bowman (2015). • 23. Barron (2010). • 24. Kringelbach (2010); Olds (1954). • 25. Sapolsky (2017), p.70. • 26. Pendergrast (2009). • 27. Hanson (2015), p.147. • 28. Tana (2015). • 29. Wild (2010), p.31. • 30. Wild (2010), p.13; Schenck (2019), p.20. • 31. Winkelman (2019), p.42; Halpern (2004); Halpern (2010). • 32. Cowan (2004). • 33. Wild (2010), p.13. • 34. Topik (2004). • 35. Bragg (2019). • 36. Bragg (2004). • 37. Benn (2005). • 38. Bragg (2004). • 39. Wild (2010), p.16. • 40. Luttinger (2006), ch.1. • 41. Pendergrast (2010), p.24. • 42. Walker (2018), loc.235. • 43. Walker (2018), loc.458; Bjorness (2009). • 44. Walker (2018), loc.465. • 45. Nathanson (1984). • 46. Wright (2013); Couvillon (2015); Stevenson (2017). • 47. Solinas (2002). • 48. Ohler (2016), ch.2; Wolfgang (2006); Doyle (2005). • 49. Pollan (2021), loc.1550; Bragg (2004). • 50. Pollan (2021), loc.1580. • 51. Walker (2018), loc.593. • 52. Öberg (2011) • 53. World Health Organisation (2021), p.17. • 54. Plants of the World Online, Royal Botanical Gardens, Kew. https://powo.science.kew.org/taxon/325974-2 • 55. Carmody (2018); Tushingham (2013); Duke (2021). • 56. Duke (2021). • 57.

Gately (2003), p.3. • 58. Gately (2003), p.14. • 59. Watson (2012), p.216; Gately (2003), p.10. • 60. Elferink (1983). • 61. Gately (2003), p.10; Elferink (1983). • 62. Mineur (2011). • 63. Charlton (2004); Mishra (2013); Goodman (1993), p.44; Gately (2003), p.4. • 64. Gately (2003), p.39. • 65. Gately (2003), p.44. • 66. Doll (1998). • 67. Gately (2003), p.23. • 68. Gately (2003), p.7. • 69. Gately (2003), p.4. • 70. Watson (2012), p.215; Gately (2003), p.8. • 71. Gateley (2003). • 72. Gately (2003), p.23. • 73. Hodge (1912), p.767. • 74. Doig (2022), p.272. • 75. Ho (2020). • 76. Biasi (2012). • 77. Gately (2003), p.37. • 78. Gately (2003), p.38; Burns (2006), p.29. • 79. Benedict (2011). • 80. Gately (2003), pp.44, 60. • 81. Burns (2006), p.43. • 82. Burns (2006), pp.50, 52. • 83. Doig (2022), p.268; Gately (2003), p.57. • 84. Woodward (2009), p.191. • 85. Gately (2003), p.70; Burns (2006), p.57. • 86. Gately (2003), p.59; Doig (2022), p.71. • 87. Sherman (2005), p.59. • 88. Gately (2003), p.72. • 89. Mann (2011), ch2. • 90. Wells (1975), p.160. • 91. Mabbett (2005); Lisuma (2020). • 92. Carr (1989). • 93. Gately (2003), p.65. • 94. Gately (2003), p.72. • 95. Milov (2019), p.2. • 96. Milov (2019), p.22. • 97. Verpoorte (2005). • 98. Verpoorte (2005). • 99. Wigner (1960). • 100. Ostlund (2017). • 101. Zimmerman (2012). • 102. Sporchia (2021). • 103. Steppuhn (2004). • 104. Morris (2011), loc.244. • 105. Bernstein (2009), loc.4965; Hanes (2002), p.20. • 106. Harrison (2017). • 107. Bernstein (2009), loc.4965. • 108. Bernstein (2009), loc.4970. • 109. Brownstein (1993); Norn (2005). • 110. Morris (2011), loc.250. • 111. Roxburgh (2020). • 112. Marr (2013), loc.7670. • 113. Morris (2011), loc.250. • 114. Bernstein (2009), loc.4970. • 115. Standage (2006), p.156. • 116. Pollan (2021), loc.1770. • 117. Paine (2015), p.522; Bernstein (2009), loc.4980. • 118. Bernstein (2009), loc.4980; Greenberg (1969), p.110. • 119. Paine (2015), p.522; Bernstein (2009), loc.5020. • 120. Morris (2011), figure 10.5. • 121. Marr (2013), loc.7690. • 122. Bernstein (2009), loc.5009; Bernstein (2009), loc.5019; Kalant (1997). • 123. Morris (2011), loc.8100. • 124. Bernstein (2009), loc.5020. • 125. Bernstein (2009), loc.5025. • 126. Bernstein (2009), loc.5025; Morris (2011), loc.8100; Hanes (2002) (2006), p.37. • 127. Hanes (2002) (2006), p.49. • 128. Marr (2013), loc.7655; Hanes (2002) (2006), p.55. • 129. Morris (2011), loc.250; 'Modern equivalent provided by Bank of England historical inflation calculator',

available at: https://www.bankofengland.co.uk/monetary-policy/infla-
tion/inflation-calculator • 130. Hanes (2002), ch4. • 131. Fay (1997),
p.261; Hanes (2002), pp.115, 199. • 132. Hanes (2002), ch.11. • 133.
Newman (1995). • 134. Zheng (2003). • 135. Hanes (2002) (2006),
p.296. • 136. United Nations Office on Drugs and Crime (2021);
United Nations Office on Drugs and Crime (2022). • 137. Centers
for Disease Control and Prevention (CDC) (2022). • 138. Health and
Human Services (2017). • 139. CDC (2022); Volkow (2021).

CHAPTER 7

1. Willyard (2018). • 2. Nachman (2000); Xue (2009). • 3. Carter
(2009). • 4. Ojeda-Thies (2003). • 5. Hibbert (2007), p.148. • 6.
Cartwright (2020), loc.3230. • 7. Arruda (2018). • 8. Cartwright
(2020), loc.3228. • 9. Ojeda-Thies (2003). • 10. Massie (1989),
p.141. • 11. Stevens (2005). • 12. Ojeda-Thies (2003); Stevens
(2005). • 13. Ojeda-Thies (2003). • 14. Ojeda-Thies (2003). • 15.
Cartwright (2020), loc.3241. • 16. Figes (1997), p.27. • 17. Stevens
(2005). • 18. Massie (1989), p.184 • 19. Fuhrmann (2012). • 20.
Stevens (2005); Fuhrmann (2012). • 21. Massie (1989), p.191. • 22.
Fuhrmann (2012). • 23. Massie (1989), p.177. • 24. Harris
(2016) • 25. Stevens (2005). • 26. Figes (1997), p.278. • 27. Figes
(1997), p2.77. • 28. Figes (1997), p.278 • 29. Figes (1997), p.33. • 30.
Figes (1997), p.33. • 31. Figes (1997), p.33. • 32. Massie (1989),
p.154. • 33. Cartwright (2020), loc.3435. • 34. Figes (1997),
p.34. • 35. Figes (1997), p.284. • 36. Massie (1989), p.217. • 37.
Stevens (2005). • 38. Cartwright (2020), loc.3445. • 39. Cartwright
(2020), loc.3500. • 40. Harris (2016). • 41. Pitre (2016). • 42.
McCord (1971). • 43. Lamb (2001), p.117; Allan (2021). • 44.
Lamb (2001), p.117. • 45. Hawkins (1847), Section XVI; Vogel
(1933). • 46. Brown (2003), p.3. • 47. Paine (2015), p.476. • 48.
Lents (2018), loc.2914. • 49. Lents (2018), loc.2920. • 50. Crittenden
(2017). • 51. Webber (2015), loc.2711; Lents (2018), loc.780. • 52.
McGee (2004), p534; Han (2021). • 53. Kluesner (2014). • 54.
Baron (2009). • 55. Linster (2006). • 56. Johnson (2010). • 57.
Lents (2018), loc.590; Nishikimi (1992); Cui (2010). • 58. Lents
(2018), loc.585. • 59. Lents (2018), loc.625. • 60. Baron
(2009). • 61. Severin (2008), p.17. • 62. Baron (2009); George
(2016). • 63. Baron (2009). • 64. Baron (2009); Vogel (1933). • 65.

Baron (2009). • 66. Baron (2009). • 67. Baron (2009). • 68. Baron (2009). • 69. Baron (2009). • 70. Vale (2008); Baron (2009). • 71. Baron (2009); Lloyd (1981). • 72. Brown (2003), p.201. • 73. Birkett (1984). • 74. Graham (1948). • 75. Duffy (1992), p.62. • 76. Graham (1948). • 77. Graham (1948). • 78. Graham (1948). • 79. Mahan (1895); Barnett (2005). • 80. Baron (2009); Lloyd (1981). • 81. Baron (2009); Lloyd (1981). • 82. Brown (2003), p.197; Loyd (1981). • 83. Brown (2003), p.195; Lloyd (1981). • 84. Southey (1813), ch.8. • 85. Allan (2021). • 86. Baron (2009). • 87. Baron (2009); Lloyd (1981). • 88. Riehn (1990), p.395. • 89. Baron (2009). • 90. Attlee (2015), p.64. • 91. Baron (2009); Carpenter (2012). • 92. Baron (2009). • 93. Watt (1981). • 94. Baron (2009). • 95. Baron (2009); Attlee (2015), p.64. • 96. Williams (1991) • 97. 'Limey, n.'. OED Online. December 2022. Oxford University Press. https://www.oed.com/view/Entry/108467 • 98. Dimico (2017). • 99. Cavaioli (2008) • 100. Rajakunmar (2003); Wheeler (2019). • 101. Schæbel (2015). • 102. Kedishvili (2017). • 103. Unicef (2021); Zhao (2022). • 104. Beyer (2002).

CHAPTER 8

1. Bernstein (2009), loc.2810. • 2. Shermer (2012), loc.4970; Kingsbury (1992). • 3. Nickerson (1998). • 4. Wrangam (2019), p.53. • 5. The White House (2005). • 6. Lents (2018), loc.2440. • 7. Lents (2018), loc.2445; Shermer (2012), loc.4576; Münchau (2017); Lerman (2018); Knobloch-Westerwick (2017); Kobloch-Westerwick (2015); Dahlgren (2019); Knobloch-Westerwick (2019). • 8. Watson (2022); Walker (2021); Ofcom (2019). • 9. Simon (1955). • 10. Miller (1956); Cowan (2010). • 11. Lents (2018), loc.2435. • 12. Kahneman (2012), p.4. • 13. Rozenkrantz (2021). • 14. Tversky (1974). • 15. Kahneman (2012), p.20. • 16. Howard (2019). • 17. Haselton (2015); Wilke (2009). • 18. Cosmides(1994); Gigerenzer (2004); Haselton (2015), p.963. • 19. Koehler (2004), p.10; Evans (2003); Kahneman (2012), p.20; Stanovich (2008). • 20. Tobin (2009) states that the phrase 'curse of knowledge' was first coined by Camerer (1989). • 21. Details of the events surrounding the Charge of the Light Brigade are provided by Brighton (2005) and David (2018). The example of the Charge of the Light Brigade as a case of miscommunication due to the curse of knowledge is given

in the report Polansky (2020), prepared for the US Department of Defence, which references a Pinker (2014) article on how this bias hampers clear science communication. This story is expanded in the excellent Harford (2021) podcast episode. • 22. Klein (2005), ch.6. • 23. Tobin (2009). • 24. For more on how groups can often make better decisions than any single member see Surowiecki (2004). On the topic of the circumstances when groups don't make better judgements, see Kahneman (2012), p.84. • 25. *The New York Times* (1973). • 26. Details on the design and development of Concorde are provided in Leyman (1986); Collard (1991); Talbort (1991); Eames (1991). • 27. Seebass (1997). • 28. Modern equivalent provided by the Inflation Calculator at www.inflationtool.com • 29. Eames (1991). • 30. Eames (1991). • 31. Dowling (2020). • 32. Dowling (2016). • 33. Teger (1980). • 34. Eames (1991). • 35. Shermer (2012), loc.4690. • 36. Dawkins (1976); Arkes (1999). • 37. Teger (1980); Schwartz (2006). • 38. BBC News (2017). • 39. The White House (2017); Owens (2021); Coy (2021). • 40. Vis (2011); Kahneman (1979); Kahneman (2012), p.302 • 41. Kahneman (1991). • 42. Lents (2018), loc.2740. • 43. Kahneman (1979). • 44. The Royal Swedish Academy of Sciences (2002). • 45. McDermott (2009); Vis (2011). • 46. Kimball (2022). • 47. McDermott (2009). • 48. Mercer (2005); Schaub (2004). • 49. Liberman (2001). • 50. McDermott (2004). • 51. Hancock (2010). • 52. Tversky (1981); Livneh (2019). • 53. Hancock (2010).

CODA

1. Swallow (2003); Ségurel (2017). • 2. Gerbault (2011). • 3. Balter (2005); Stock (2008). • 4. Cregan-Reid (2018), p.168; Pan (2011); Holden (2016). • 5. Rao (2018); Blumenthal (2010). • 6. Corner (2012). • 7. References for this section on cognitive biases and the challenges of climate change: Clayton (2015); Zaval (2016). King (2019); Zhao (2021); Moser (2021).

Bibliography

Acemoglu, D., Johnson, S. and Robinson, J. A. (2001). 'The colonial origins of comparative development: an empirical investigation'. *American Economic Review*, 91 (5), 1369–1401.

Acevedo, B. P. and Aron, A. P. (2014). 'Romantic love, pair-bonding, and the dopaminergic reward system'. *American Psychological Association*, 55–69.

Achan, J., Talisuna, A. O., Erhart, A., Yeka, A., Tibenderana, J. K., Baliraine, F. N., Rosenthal, P. J. and D'Alessandro, U. (2011). 'Quinine, an old anti-malarial drug in a modern world: role in the treatment of malaria'. *Malaria Journal*, 10 (144).

Akinyanju, O. O. (1989). 'A profile of sickle cell disease in Nigeria'. *Annals of the New York Academy of Sciences*, 565, 126–136.

Alexander, R. D. (2020). 'The Biology of Moral Systems'. *Canadian Journal of Philosophy*, 21 (2).

Alfani, G. (2013). 'Plague in seventeenth-century Europe and the decline of Italy: an epidemiological hypothesis'. *European Review of Economic History*, 17 (4), 408–430.

Alfani, G. and Murphy, T. E. (2017). 'Plague and lethal epidemics in the pre-industrial world'. *Journal of Economic History*, 77 (1).

Allan, P. K. (2021). 'Finding a cure for scurvy'. *Naval History Magazine*, 35 (1).

Alvarez, G., Ceballos, F. C. and Quinteiro, C. (2009). 'The role of inbreeding in the extinction of a European royal dynasty'. *PLoS One*, 4 (4).

Anderson, J., and Shuttleworth, I. (1998). 'Sectarian demography, territoriality and political development in Northern Ireland'. *Political Geography*, 17 (2), 187–208.

Anter, A. (2019). 'The Modern State and Its Monopoly on Violence'. In: Hanke, E., Scaff, L. and Whimster, S. (eds).*The Oxford Handbook of Max Weber*. Oxford University Press.

Archarya, A. and Lee, A. (2019). 'Path dependence in European development: medieval politics, conflict and state building'. *Comparative Political Studies*, 52 (13).

Arkes, H. R. and Ayton, P. (1999). 'The sunk cost and Concorde effects: are humans less rational than lower animals?' *Psychological Bulletin*, 125, 591–600.

Armelagos, G. J., Goodman A. H. and Jacobs, K. H. (1991). 'The Origins of Agriculture: Population Growth during a Period of Declining Health'. *Population and Environment*, 13 (1), 9–22.

Armitage, D. (1994). 'The projecting age: William Paterson and the Bank of England'. *History Today*, 44 (6).

Arnold, D. (2019). 'Death and the modern empire: the 1918–19 influenza epidemic in India'. *Transactions of the Royal Historical Society*, 29, 181–200.

Arruda, V. R., and High, K. A. (2018). 'Coagulation disorders'. In: Jameson, J., Fauci, A. S., Kasper, D. L., Hauser, S. L., Longo, D. L., and Loscalzo, J. (eds). *Harrison's Principles of Internal Medicine*, 20e. McGraw Hill.

Atran, S. (2001). 'A cheater-detection module?' *Evaluation and Cognition*, 7 (2), 1–7.

Attlee, H. (2015). *The land Where Lemons Grow: the story of Italy and its citrus fruit*. Penguin.

Ayres, L. P. (1919). *The war with Germany: a statistical summary*. Washington Government Printing Press. Available at: https://archive.org/details/warwithgermanystooayreuoft

Badiaga, S. and Brouqui, P. (2012). 'Human louse-transmitted infectious diseases'. *Clinical Microbiology and Infection*, 18 (4), 332–337.

Balter, M. (2005). 'Are humans still evolving?' *Science*, 309 (5732), 234–237.

Bamford, S. (2019). *The Cambridge Handbook of Kinship*. Cambridge University Press.

Barboza Retana, F. A. (2002). 'Two Discoveries, Two Conquests, and Two Vázquez de Coronado'. *Diálogos Revista Electrónica de*

Historia, 3 (2–3). Available at: https://www.redalyc.org/articulo. oa?id=43932301

Barnett, R. W. (2005). 'Technology and Naval Blockade: Past Impact and Future Prospects'. *Naval War College Review*, 58(3), 87–98.

Baron, J. H. (2009). 'Sailors' scurvy before and after James Lind – a reassessment'. *Nutrition Reviews*, 67 (6), 315–332.

Barron, A. B., Søvik, E. and Cornish, J. L. (2010). 'The Roles of Dopamine and Related Compounds in Reward-Seeking Behavior Across Animal Phyla'. *Frontiers in Behavioral Neuroscience*, 4, 163.

Bartlett, R. (2020). *The James Lydon Lectures in Medieval History and Culture*. Cambridge University Press.

BBC News (2017). 'US sends 3,000 more troops to Afghanistan'. BBC News, 18 September 2017. https://www.bbc.co.uk/news/ world-us-canada-41314428

BBC News (2022). 'NI election results 2022: Sinn Féin wins most seats in historic election'. BBC News, 8 May 2022. https://www. bbc.co.uk/news/uk-northern-ireland-61355419

Beare, W. (1964). 'Tacitus on the Germans'. *Greece & Rome*, 11 (1), 64–76.

Beckert, J. (2007). 'The "long durée" of inheritance law: discourses and institutional development in France, Germany, and the United States since 1800'. *European Journal of Sociology*, 48 (1), 79–120.

Behrendt, L. (2010). 'Consent in a (Neo)Colonial Society: Aboriginal Women as Sexual and Legal "Other"'. *Australian Feminist Studies*, 15 (33), 353–367.

Benedict, C. (2011). *Golden-Silk Smoke: A History of Tobacco in China, 1550–2010*. University of California Press.

Benn, J. A. (2005). 'Buddhism, Alcohol, and Tea in Medieval China'. In: Sterckx, R. (ed.). *Of Tripod and Palate: Food, Politics, and Religion in Traditional China*. Palgrave Macmillan.

Berniell-Lee, G., Calafell, F., Bosch, E., Heyer, E., Sica, L., Mouguiama-Daouda, P., van der Veen, L., Hombert, J. M., Quintana-Murci, L. and Comas, D. (2009). 'Genetic and demographic implications of the Bantu expansion: insights from human paternal lineages'. *Molecular Biology and Evolution*, 26 (7), 1581–1589.

Bernstein, W. L. (2009). *A splendid exchange: how trade shaped the world*. Atlantic Books.

Bertocchi, G. and Dimico, A. (2015). 'The long-term determinants of female HIV infection in Africa: the slave trade, polygyny, and sexual behaviour'. *Journal of Development Economics*, 140, 90–105.

Bethmann, D. and Kvasnicka, M. (2012). 'World War II, missing men and out of wedlock childbearing'. *Economic Journal*, 123 (567), 162–194.

Bettinger, R. L. (2016). 'Prehistoric hunter-gatherer population growth rates rival those of agriculturalists'. *Proceedings of the National Academy of Sciences*, 113 (4), 812–814.

Betzig, L. (2014). 'Eusociality in history'. *Human Nature*, 25, 80–99.

Beyer, P., Al-Babili, S., Ye, X., Lucca, P., Schaub, P., Welsch, R. and Potrykus, I. (2002). 'Golden rice: introducing the b-carotene biosynthsis pathway into rice endosperm by genetic engineering to defeat vitamin A deficiency'. *Journal of Nutrition*, 132 (3), 506–510.

Biasi, M. D. and Dani, J. A. (2012). 'Reward, addiction, withdrawal to nicotine'. *Annual Review of Neuroscience*, 34, 105–130.

Birkett, J. D. (1984). 'A brief illustrated history of desalination: from the Bible to 1940'. *Desalination*, 50, 17–52.

Bixler, R. H. (1982). 'Sibling incest in the royal families of Egypt, Peru and Hawaii'. *Journal of Sex Research*, 18 (3), 264–281.

Bjorness, T. E. and Greene, R. W. (2009). 'Adenosine and sleep'. *Current Neuropharmacology*, 7 (3), 238–245.

Blanc, G. (2021). 'Modernization Before Industrialization: Cultural Roots of the Demographic Transition in France'. Working paper, available at: http://dx.doi.org/10.2139/ssrn.3702670

Blumenthal, D. M., Gold, M. S. (2010). 'Neurobiology of food addiction'. *Current Opinion in Clinical Nutrition and Metabolic Care*, 13 (4), 359–365.

Bostoen, K. (2018). 'The Bantu Expansion'. *Oxford Research Encyclopedia of African History*. Oxford University Press.

Bostoen, K. (2020). 'The Bantu Expansion: Some facts and fiction'. In: Crevels, M. and Muysken, P. (eds). *Language Dispersal, Diversification, and Contact*. Oxford University Press.

Bostwick, W. (2015). 'How the India Pale Ale Got Its Name'. *Smithsonian Magazine*. Available at: https://www.smithsonianmag.com/history/how-india-pale-ale-got-its-name-180954891/

Bowles, S. (2011). 'Cultivation of cereals by the first farmers was not more productive than foraging'. *Proceedings of the National Academy of Sciences of the United States of America*, 108 (12), 4760–4765.

Bowman, E. (2015). 'Explainer: what is dopamine – and is it to blame for our addictions?' *The Conversation*. Available at: https://the conversation.com/explainer-what-is-dopamine-and-is-it-to-blame-for-our-addictions-51268.

Boyd, R., Gintis, H. and Bowles, S. (2010). 'Coordinated punishment of defectors sustains cooperation and can proliferate when rare'. *American Association for the Advancement of Science*, 328 (5978), 617–620.

Boyd, R., Gintis, H., Bowles, S. and Richerson, P.J. (2003). 'The evolution of altruistic punishment'. *Proceedings of the National Academy of Sciences of the United States of America*, 100 (6), 3531–3535.

Bragg, M. (2004). *Tea. In Our Time*, BBC Radio 4. Available at: https://www.bbc.co.uk/programmes/p004y24y

Bragg, M. (2019). *Coffee. In Our Time*, BBC Radio 4. Available at: https://www.bbc.co.uk/programmes/m000c4x1

Brainerd, E. (2017). 'The lasting effect of sex ratio imbalance on marriage and family: evidence from World War II in Russia'. *The Review of Economics and Statistics*, 99 (2), 229–242.

Brewer, H. (1997). 'Entailing aristocracy in colonial Virginia: "ancient feudal restraints" and revolutionary reform'. *Omohundro Institute of Early American History and Culture*, 54 (2), 307–346.

Brighton, T. (2005). *Hell Riders: the truth about the Charge of the Light Brigade*. Penguin.

Brook, T. (2013). *The Troubled Empire: China in the Yuan and Ming dynasties*. Harvard University Press.

Brooks, P. J., Enoch, M. A., Goldman, D., Li, T. K. and Yokoyama, A. (2009). 'The Alcohol Flushing Response: An Unrecognized Risk Factor for Esophageal Cancer from Alcohol Consumption'. *PLoS Medicine*, 6 (3).

Brown, S. P. (2003). *Scurvy: How a Surgeon, a Mariner, and a Gentleman Solved the Greatest Medical Mystery of the Age of Sail*. Thomas Dunne Books.

Brownstein, M. J. (1993). 'A brief history of opiates, opioid peptides, and opioid receptors'. *Proceedings of the National Academy of Sciences*, 90 (12), 5391–5393.

Bryant, J. E., Holmes, E. C. and Barrett, A.D.T. (2007). 'Out of Africa: a molecular perspective on the introduction of yellow fever virus into the Americas'. *PLoS Pathogens*, 3 (5).

Burns, E. (2006). *The Smoke of the Gods: a social history of tobacco.* Temple University Press.

Burton-Chellew, M. N. and Dunbar, R.I.M. (2015). 'Hamilton's rule predicts anticipated social support in humans'. *Behavioral Ecology,* 26 (1), 130–137.

Bush, R. D. (2013). *The Louisiana Purchase: a global context.* Taylor & Francis Group.

Buss, D. M. (ed.) (2015). *The Handbook of Evolutionary Psychology.* John Wiley & Sons Inc.

Camerer, C. F., Loewenstein, G. F. and Weber, M. (1989). 'The Curse of Knowledge in Economic Settings: An Experimental Analysis'. *Journal of Marketing,* 53(5), 1–20.

Campbell, L. and Ellis, B. J. (2015). 'Commitment, Love, and Mate Retention'. In: Buss, D. M. (ed.). *The Handbook of Evolutionary Psychology.* Wiley.

Campbell-Platt, G. (1994). 'Fermented foods – a world perspective'. *Food Research International,* 27 (3), 253–257.

Carlisle, E. and Shafir, E. (2002). 'Questioning the cheater-detection hypothesis: New studies with the selection task'. *Thinking and Reasoning,* 11 (2), 97–122.

Carmody, S.B., Davis, J., Tadi, S., Sharp, J., Hunt, R. and Russ, J. (2018). 'Evidence of tobacco from a Late Archaic smoking tube recovered from the Flint River site in southeastern North America'. *Journal of Archaeological Science,* 21, 904–910.

Carpenter, K. J. (2012). 'The Discovery of Vitamin C'. *Annals of Nutrition and Metabolism,* 61, 259–264.

Carr, L. G. and Menard, R. R. (1989). 'Land, labor, and economies of scale in early Maryland: some limits to growth in the Chesapeake system of husbandry'. *Journal of Economic History,* 49 (2), 407–418.

Carrigan, M. A., Uryasev, O., Frye, C. B., Eckman, B. L., Myers, C. R., Hurley, T. D. and Benner, S. A. (2014). 'Hominids adapted to metabolize ethanol long before human-directed fermentation'. *Proceedings of the National Academy of Sciences of the United States of America,* 112 (2), 458–463.

Carroll, R. (2007). 'The sorry story of how Scotland lost its 17th-century empire'. *Guardian,* 11 September 2007. Available at: https://www.theguardian.com/uk/2007/sep/11/britishidentity.past

Carroll, R., O'Carroll, L. and Helm, T. (2022). 'Sinn Féin assembly victory fuels debate on future of union'. *Observer*, 8 May 2022. https://www.theguardian.com/politics/2022/may/07/ sinn-fein-assembly-victory-fuels-debate-on-future-of-union

Carron, P. M., Kaski, K. and Dunbar, R. (2016). 'Calling Dunbar's numbers'. *Social Networks*, 47, 151–155.

Carter, M. (2009). 'The last emperors'. *Guardian*, 12 Sep 2009. Available at: https://www.theguardian.com/lifeandstyle/2009/ sep/12/queen-victoria-royal-family-europe

Cartwright, F. F. and Biddiss, M. (2020). *Disease and History: From Ancient Times to Covid-19*, 4th edition. Lume Books.

Cartwright, J. (2000). *Evolution and Human Behavior: Darwinian perspectives on human nature*. MIT Press.

Cavaioli, F. J. (2008). 'Patterns of Italian Immigration to the United States'. *Catholic Social Science Review*, 13, 213–229.

Cavalli-Sforza, L. L., Cavalli-Sforza, L., Menozzi, P., Piazza, A. (1994). *The History and Geography of Human Genes*. Princeton University Press.

Centers for Disease Control and Prevention (CDC) (2022). 'Understanding the Opioid Overdose Epidemic'. Available at: https://www.cdc.gov/opioids/basics/epidemic.html

Central Intelligence Agency (2021). *The World Factbook 2021*. Washington, DC. Available at: https://www.cia.gov/the-world-factbook/field/sex-ratio

Chamorro-Premuzic, T. (2015). 'Reputation and the rise of the rating society'. *Guardian*, 26 October 2015. Available at: https://www. theguardian.com/media-network/2015/oct/26/ reputation-rating-society-uber-airbnb

Chandra, S. and Kassens-Noor, E. (2014). 'The evolution of pandemic influenza: evidence from India, 1918–19'. *BMC Infectious Diseases*, 14, 510.

Chandra, S., Juljanin, G. and Wray, J. (2012). 'Mortality from the influenza pandemic of 1918–1919: the case of India'. *Demography*, 49 (3), 857–865.

Charlton, A. (2004). 'Medicinal uses of tobacco in history'. *Journal of the Royal Society of Medicine*, 97 (6), 292–296.

Chittka L. and Peng, F. (2013). 'Caffeine boosts bees' memories'. *Science*, 339 (6124), 1157–1159.

Christakis, N. A. (2019). *Blueprint: the evolutionary origins of a good society*. Little, Brown and Company.

Christian, B. and Griffiths, T. (2016). *Algorithms to Live By: the computer science of human decisions*. Henry Holt and Company.

Chunn, M. (2015). *Death and Disorder: The 1918–1919 Influenza Pandemic in British India*. University of Colorado at Boulder.

Ciment, J. (2007). *Atlas of African-American History*. Infobase Publishing.

Clark, D.P. (2010). *Germs, Genes, & Civilization: how epidemics shaped who we are today*. Pearson.

Clark, A. and Alter, G. (2010). 'The demographic transition and human capital'. In: Broadberry, S. and O'Rourke, K. H. (eds). *The Cambridge Economic History of Modern Europe*. Cambridge University Press.

Clayton, S., Devine-Wright, P., Stern, P. C., Whitmarsh, L., Carrico, A., Steg, L., Swim, J. and Bonnes, M. (2015). 'Psychological research and global climate change'. *Nature Climate Change*, 5 (7), 640–646.

Clodfeiter, M. (2008). *Warfare and Armed Conflicts: a statistical encyclopedia of casualty and other figures, 1494–2007*. McFarland.

Coe, M. D. (2008). *Mexico: From the Olmecs to the Aztecs*. Thames & Hudson.

Collard, D. (1991). 'Concorde airframe design and development'. Journal of *Aerospace*, 100, 2620–2641.

Collins, L. (2006). 'Choke Artist'. *New Yorker*, 8 May 2006. Available at: https://www.newyorker.com/magazine/2006/05/08/choke-artist

Compton, P.A. (1976). 'Religious Affiliation and Demographic Variability in Northern Ireland'. *Transactions of the Institute of British Geographers*, 1 (4), 433–452.

Corner, A., Whitmarsh, L. and Xenias, D. (2012). 'Uncertainty, scepticism and attitudes towards climate change: biased assimilation and attitude polarisation'. *Climatic Change*, 114, 463–478.

Cosmides, L. (1989). 'The logic of social exchange: has natural selection shaped how humans reason? Studies with the Wason selection task'. *Cognition*, 31, 187–276.

Cosmides, L., Barrett, C. and Tooby, J. (2010). 'Adaptive specializations, social exchange, and the evolution of human intelligence'. *Proceedings of the National Academy of Sciences of the United States of America*, 107 (2), 9007–9014.

Cosmides, L., Tooby, J. (1994). 'Better than Rational: Evolutionary Psychology and the Invisible Hand'. *American Economic Review*, 84(2), 327–332.

Cosmides, L. and Tooby, J. (2015). 'Neurocognitive Adaptations Designed for Social Exchange'. In: Buss, D. M. (ed.). *The Handbook of Evolutionary Psychology*. Wiley.

Couvillon, M. J., Al Toufailia, H., Butterfield, T. M., Schrell, F., Ratnieks, F.L.W. and Schürch, R. (2015). 'Caffeinated forage tricks honeybees into increasing foraging and recruitment behaviors'. *Current Biology*, 25 (21), 2815–2818.

Cowan, B. (2004). 'The rise of the coffeehouse reconsidered'. *Historical Journal*, 47 (1), 21–46.

Cowan, N. (2010). 'The magical mystery four: how is working memory capacity limited, and why?' *Current Directions in Physiological Science*, 19 (1).

Coy, P. (2021). 'America's War in Afghanistan Is the Mother of All Sunk Costs'. Bloomberg UK, 19 April 2021. https://www.bloomberg. com/news/articles/2021-04-19/america-s-war-in-afghanistan-is-the-mother-of-all-sunk-costs

Crawford, D. H. (2009). *Deadly companions: how microbes shaped our history*. Oxford University Press.

Cregan-Reid, V. (2018). *Primate Change: How the world we made is remaking us*. Octopus Books .

Crittenden, A. N. and Schnorr, S. L. (2017). 'Current views on hunter-gatherer nutrition and the evolution of the human diet'. *American Journal of Physical Anthropology*, 162 (63), 84–109.

Cui, J., Pan, Y. H., Zhang, Y., Jones, G. and Zhang, S. (2010). 'Progressive pseudogenization: vitamin C synthesis and its loss in bats'. *Molecular Biology and Evolution*, 28 (2), 1025–1031.

Cummins, N. (2009). 'Marital fertility and wealth in transition era France, 1750–1850'. Working Paper No. 2009–16. Paris School of Economics.

Dahlgren, P. M., Shehata, A. and Strömbäck, J. (2019). 'Reinforcing spirals at work? Mutual influences between selective news exposure and ideological leaning'. *European Journal of Communication*, 34 (2).

Dale, M. S. (2017). 'Running Away from the Palace: Chinese Eunuchs during the Qing Dynasty'. *Journal of the Royal Asiatic Society*, 27(1), 143–164.

Dale, M. S. (2018). *Inside the World of the Eunuch: A Social History of the Emperor's Servants in Qing China*. Hong Kong University Press.

Dalton, J. T. and Leung, T. C. (2014). 'Why is polygyny more prevalent in Western Africa? An African slave trade perspective'. *Economic Development and Cultural Change*, 62 (4).

Dapa, D. and Gil, T. (2002). 'Sickle cell disease in Africa'. *Erythrocytes*, 9 (2), 111–116.

Darwin, C. (1839). *The voyage of the Beagle*. Available at: https://www.gutenberg.org/files/944/944-h/944-h.htm

Darwin, C. (1859). *On the Origin of Species*. Available at: https://www.gutenberg.org/files/1228/1228-h/1228-h.htm

David, S. (2018). 'The Charge of the Light Brigade: who blundered in the Valley of Death?' *History Extra*. Available at: https://www.history extra.com/period/victorian/the-charge-of-the-light-brigade-who-blundered-in-the-valley-of-death/

Dawkins, R. and Carlisle, T. R. (1976). 'Parental investment, mate desertion and a fallacy'. *Nature*, 262, 131–133.

de Filippo, C., Bostoen, K., Stoneking, M. and Pakendorf, B. (2012). 'Bringing together linguistic and genetic evidence to test the Bantu expansion'. *Proceedings of the Royal Society B*, 279, 3256–3263.

de Barros, J. (1552) Decada Primeira, Livro 3. Translated sections available in Chapter 2 of: Boxer, C.R. (1969) Four Centuries of Portuguese Expansion, 1415-1825: A Succinct Survey. Witwatersrand University Press.

de Luna, K. M. (2018). 'Language Movement and Change in Central Africa'. In: Albaugh, E. A. and de Luna, K. M. (eds). *Tracing Language Movement in Africa*. Oxford University Press.

de Montaigne, M. (1580) 'Of Friendship'. In: Hazilitt, W. C. (Ed.) (1877) *Essays of Michel de Montaigne*. Translated by Charles Cotton. Available from: https://www.gutenberg.org/cache/epub/3586/pg3586.html

de Quervain, D.J.F., Fischbacher, U., Treyer, V., Schellhammer, M., Schnyder, U., Buck, A. and Fehr, E. (2004). 'The neural basis of altruistic punishment'. *Science*, 305 (5688), 1254–1258.

de Waal, F.B.M. (1997). 'The chimpanzee's service economy: Food for grooming'. *Evolution and Human Behavior*, 18, 375–386.

Dean, M., Carrington, M., Winkler, C., Huttley, G. A., Smith, M. W., Allikmets, R. (1996). 'Genetic restriction of HIV-1 infection and progression to AIDS by a deletion allele of the CKR5 structural gene'. *Science*, 273(5283), 1856–62.

Denevan, W. M. (2010). 'The pristine myth: the landscape of the Americas in 1942'. *Annals of the Association of American Geographers*, 369–385.

Depetris-Chauvin, E. and Weil, D. N. (2013). 'Malaria and early African development: evidence from the sickle cell trait'. *Economic Journal* 128 (610), 1207–1234.

Desan, S. (1997). '"War between Brothers and Sisters": Inheritance Law and Gender Politics in Revolutionary France'. *French Historical Studies*, 20 (4), 597–634.

Diamond, J. (1987). 'The Worst Mistake in the History of the Human Race'. *Discover Magazine*, May 1, 1987. Pages 64–66.

Diamond, J. (1998). *Guns, Germs and Steel: A short history of everybody for the last 13,000 years*. Vintage.

Diamond, J. and Bellwood, P. (2003). 'Farmers and their languages: the first expansions'. *Science*, 300 (5619), 597–603.

Diamond, J. and Robinson, J. A. (eds) (2011). *Natural Experiments of History*. Harvard University Press.

Dimico, A., Isopi, A. and Olsson, O. (2017). 'Origins in the Sicilian Mafia: the market for lemons'. *Journal of Economic History*, 77 (4), 1083–1115.

Dobson, M. J. (2007). *Disease: The Extraordinary Stories Behind History's Deadliest Killers*. Quercus.

Dobyns, H. F. (1966). 'An appraisal of techniques with a new hemispheric estimate'. *Current Anthropology*, 7 (4).

Doig, A. (2022). *This Mortal Coil: A History of Death*. Bloomsbury Publishing.

Dolivo, V. and Taborsky, M. (2015). 'Norway rats reciprocate help according to the quality of help they received'. *Biology Letters*, 11(2).

Doll, R. (1998). 'Uncovering the effects of smoking: historical perspective'. *Statistical Methods in Medical Research*, 7, 87–117.

Dominy, N. J. (2015). 'Ferment in the family tree'. *Proceedings of the National Academy of Sciences of the United States of America*, 112 (2), 308–309.

Doolan, D.L., Dobaño, C. and Kevin Baird, J. (2009). 'Acquired immunity to malaria'. *Clinical Microbiology Reviews*, 22 (1), 13–36.

Dounias, E. (2001). 'The management of wild yam tubers by the Baka pygmies in southern Cameroon'. *African Study Monographs*, suppl. 26,135–156.

Dove, T. (2021). 'How Sickle Cell Trait in Black People Can Give the Police Cover'. *New York Times*, 15 May 2021. https://www.nytimes.com/2021/05/15/us/african-americans-sickle-cell-police.html

Dowling, S. (2016). 'The American Concordes that never flew'. BBC Future. Available at: https://www.bbc.com/future/article/20160321-the-american-concordes-that-never-flew.

Dowling, S. (2020). 'The Soviet Union's flawed rival to Concorde'. BBC Future. https://www.bbc.com/future/article/20171018-the-soviet-unions-flawed-rival-to-concorde

Doyle, D. (2005). 'Adolf Hitler's medical care'. *Journal of the Royal College of Physicians of Edinburgh*, 35, 75–82.

Duffy, B. (2018). *The Perils of Perception: Why we're wrong about nearly everything*. Atlantic Books.

Duffy, M. (1992). *The Establishment of the Western Squadron as the Linchpin of British Naval Strategy. Parameters of British Naval Power, 1650–1850*. University of Exeter Press.

Dugatkin, L. A. (2007). 'Inclusive fitness theory from Darwin to Hamilton'. *Genetics*, 3 (1), 1375–1380.

Duindam, J. (2015). *Dynasties: A Global History of Power, 1300–1800*. Cambridge University Press.

Duindam, J. (2019). *Dynasty: A Very Short Introduction*. Oxford University Press.

Duke, D., Wohlgemuth, E., Adams, K. R., Armstrong-Ingram, A., Rice, S. K. and Young, D. C. (2021). 'Earliest evidence for human use of tobacco in the Pleistocene Americas'. *Nature Human Behaviour*, 6, 183–192.

Dunbar, R.I.M. (1992). 'Neocortex size as a constraint on group size in primates'. *Journal of Human Evolution*, 22 (6), 469–493.

Dunbar, R.I.M., Arnaboldi, V., Conti, M. and Passarella, A. (2015). 'The structure of online social networks mirrors those in the offline world'. *Social Networks*, 43, 39–47.

Dyble, M., Thorley, J., Page, A. E., Smith, D. and Migliano, A. B. (2019). 'Engagement in agricultural work is associated with reduced leisure time among Agta hunter-gatherers'. *Nature Human Behaviour*, 3, 792–796.

Dyson, S. M. and Boswell, G. R. (2006). 'Sickle cell anaemia and deaths in custody in the UK and the USA'. *Howard Journal of Crime and Justice*, 45 (1), 14–28.

Eames, J. D. (1991). 'Concorde Operations'. *Journal of Aerospace*, 100 (1), 2603–2619.

Economist Intelligence (2022). *Democracy Index 2021: the China challenge*. Available from: https://www.eiu.com/n/campaigns/democracy-index-2021/

Edenberg, H. J. and McClintick, J. N. (2018). 'Alcohol Dehydrogenases, Aldehyde Dehydrogenases, and Alcohol Use Disorders: A Critical Review'. *Alcoholism: Clinical and Experimental Research*, 42 (12), 2281–2297.

Edlund, L. and Ku, H. (2011). 'The African Slave Trade and the Curious Case of General Polygyny'. MPRA Paper 52735, University Library of Munich, Germany.

Edwardes, M.P.J. (2019). *The Origins of Self: An Anthropological Perspective*. UCL Press.

Ehret, C. (2016). *The Civilizations of Africa: A History to 1800*. University of Virginia Press.

Eisenburg, M. and Mordechai, L. (2019). 'The Justinianic Plague: an interdisciplinary review'. *Byzantine and Modern Greek Studies*, 43(2), 156–80.

Elferink, J.G.R. (1983). 'The narcotic and hallucinogenic use of tobacco in Pre-Columbian Central America'. *Journal of Ethnopharmacology*, 7, 111–122.

Elias, S. A., Short, S. K., Nelson, C. H. and Birks, H. H. (1996). 'Life and times of the Bering land bridge'. *Nature*, 382, 60–63.

Ellman, M. and Maksudov, S. (1994). 'Soviet deaths in the great patriotic war: A note'. *Europe-Asia Studies*, 46 (4), 671–680.

Ellner, P. D. (1998). 'Smallpox: gone but not forgotten'. *Infection*, 26 (5), 263–269.

Engen, R. (2011). 'S.L.A. Marshall and the Ratio of Fire: History, Interpretation, and the Canadian Experience'. *Canadian Military History*, 20 (4), 39–48.

Esposito, E. (2015). 'Side Effects of Immunities: the African Slave Trade'. EUI Working Paper MWP 2015/09. European University Institute.

Evans, J. (2003). 'In two minds: dual-process accounts of reasoning'. *Trends in Cognitive Science*, 7 (10), 454–459.

Ewald, P. W. (1991). 'Transmission modes and the evolution of virulence with special reference to cholera, influenza and AIDS'. *Human Nature*, 2 (1), 1–30.

Falkner, J. (2021). *The War of the Spanish Succession 1701–1714*. Pen & Sword Military.

Fay, P. W. (1997). *The Opium War, 1840–1842*. The University of North Carolina Press.

Fehr, E. and Gächter, S. (2002). 'Altruistic punishment in humans'. *Nature*, 415, 137–140.

Fernández-Armesto (2019). *Out of our Minds: What we think and how we came to think it.* Oneworld Publications.

Fernández, R., Fogli, A. and Olivetti, C. (2004). 'Mothers and sons: preference formation and female labor force dynamics'. *Quarterly Journal of Economics*, 119 (4), 1249–1299.

Figes, O. (1997). *A People's Tragedy: A History of the Russian Revolution.* Viking.

Fisher, H. E. (1989). 'Evolution of human serial pairbonding'. *American Journal of Physical Anthropology*, 78 (3), 331–354.

Fisher, H. E., Aron, A. and Brown, L. L. (2006). 'Romantic love: a mammalian brain system for mate choice'. *Philosophical Transactions of the Royal Society B*, 361 (1476).

Flinn, M. V., Ward, C. V., and Noone, R. J. (2015). 'Hormones and the Human Family'. In: Buss, D. M. (ed.). *The Handbook of Evolutionary Psychology.* Wiley.

Foley, M. and Tilley, L. (1997). 'Quinoline antimalarials: Mechanisms of action and resistance'. *International Journal for Parasitology*, 27 (2), 231–240.

Fortunato, L. (2012). 'The evolution of matrilineal kinship organisation'. *Proceedings of the Royal Society B*, 279 (1749).

Fowler, J. H. (2005). 'Altruistic punishment and the origin of cooperation'. *Proceedings of the National Academy of Sciences of the United States of America*, 102 (19), 7047–7049.

Fuchs, B., Sornette, D. and Thurner, S. (2014). 'Fractal multi-level organisation of human groups in a virtual world'. *Scientific Reports*, 6526.

Fuhrmann, J. T. (2012). *Rasputin: the untold story.* Wiley.

Galdikas, B.M.F. and Wood, J. W. (1990). 'Birth spacing patterns in humans and apes'. *American Journal of Physical Anthropology*, 83 (2), 185–191.

Galvani, A. P. and Novembre, J. (2005). 'The evolutionary history of the CCR5-Δ32 HIV-resistance mutation'. *Microbes and Infection*, 7 (2), 302–309.

Gao, G. (2015). 'Why the former USSR has far fewer men than women'. Pew Research Center. Available at: https://www.pewresearch.org/fact-tank/2015/08/14/
why-the-former-ussr-has-far-fewer-men-than-women/

Gartzke, E. (2011). 'Blame it on the weather: Seasonality in Interstate Conflict'. Working paper. Available at: https://pages.ucsd.edu/~egartzke/papers/seasonality_of_conflict_102011.pdf

Gasquet, F. A. (1893). *The Great Pestilence (A.D. 1348–9), now commonly known as The Black Death*. Simpkin Marshal. Available at: https://www.gutenberg.org/files/45815/45815-h/45815-h.htm

Gately, I. (2003). *Tobacco: a cultural history of how an exotic plant seduced civilization*. Grove Press.

Gates, D. (2003). *The Napoleonic wars 1803–1815*. Pimlico.

George, A. (2016). 'How the British defeated Napoleon with citrus fruit'. *The Conversation*. Available at: https://theconversation.com/how-the-british-defeated-napoleon-with-citrus-fruit-58826

Gerbault, P., Liebert, A., Itan, Y., Powell, A., Currat, M., Burger, J., Swallow, D. M. and Thomas, M. G. (2011). 'Evolution of lactase persistence: an example of human niche construction'. *Philosophical Transactions of the Royal Society B*, 366 (1566).

Gianchecchi, E., Cianchi, V., Torelli, A. and Montomoli, E. (2022). 'Yellow fever: origin, epidemiology, preventive strategies and future prospects'. *Vaccines*, 10 (3), 372.

Gigerenzer, G. (2004). 'Fast and Frugal Heuristics: The Tools of Bounded Rationality'. In: Koehler, D. J., Harvey, N. (eds). *Blackwell Handbook of Judgement and Decision Making*. Blackwell.

Girard, P. R. (2011). *The slaves who defeated Napoleon*. University Alabama Press.

Glantz, D. M. (2005). *Colossus Reborn: The Red Army at War, 1941–1943*. University Press of Kansas.

Glass, R. I., Holmgren, J., Haley, C. E., Khan, M. R., Svennerholm, A. M., Stoll, B. J. (1985). 'Predisposition for cholera of individuals with O blood group. Possible evolutionary significance'. *American Journal of Epidemiology*, 121(6), 791–796.

Glenn, R. W. (2000). *Reading Athena's Dance Card: Men Against Fire in Vietnam*. Naval Institute Press.

Gobel, T., Waters, M. R. and O'Rourke, D. H. (2008). 'The late Pleistocene dispersal of modern humans in the Americas'. *Science*, 319 (5869), 1497–1502.

Godfrey, B. and Williams, L. (2018). 'Australia's last living convict bucked the trend of reoffending'. ABC News, 10 January 2018. Available at: https://www.abc.net.au/news/2018-01-10/australias-last-convicts/9317172

Gong, L., Parikh, S., Rosenthal, P. J. and Greenhouse, B. (2013). 'Biochemical and immunological mechanisms by which sickle cell trait protects against malaria'. *Malaria Journal*, 12 (317).

Goodman, J. (1993). *Tobacco in History: The cultures of dependence.* Routledge.

Gordon, G. (2018). 'Catholic majority possible in NI by 2021'. BBC News, 19 April 2018. Available at: https://www.bbc.co.uk/news/uk-northern-ireland-43823506

Gould, E. A., de Lamballerie, X., de A Zanotto, P. M. and Holmes, E. C. (2003). 'Origins, evolution and vector/host coadaptations within the genus Flavivirus'. *Advances in Virus Research*, 59, 277–314.

Graham, G. S. (1948). 'The naval defence of British North America 1749–1763'. *Transactions of the Royal Historical Society*, 30, 95–110.

Grange, Z. L., Goldstein, T., Johnson, C. K., Anthony, S., Gilardi, K., Daszak, P., Olival, K. J., Murray, S., Olson, S. H., Togami, E., Vidal, G. and Mazer, J. A. (2021). 'Ranking the risk of animal-to-human spillover for newly discovered viruses'. *Proceedings of the National Academy of Sciences of the United States of America*, 118 (15).

Green, E. (2013). 'Explaining African ethnic diversity'. *International Political Science Review*, 34(3), 235–253.

Green, M. H. (2017). 'The globalisations of disease'. In: Boivin, N., Crassard, R., Petraglia, M. (eds). *Human Dispersal and Species Movement: From Prehistory to the Present.* Cambridge University Press.

Greenberg, M. (1969). *British Trade and the Opening of China.* Cambridge University Press.

Greif, A. (1989). 'Reputation and coalitions in medieval trade: evidence on the Magribi traders'. *Journal of Economic History*, 49 (4), 857–882.

Grigg, D. B. (1980). *Population growth and agrarian change.* Cambridge University Press.

Grosjean, P. and Khattar, R. (2019). 'It's raining men! Hallelujah? The long-run consequences of male-biased sex ratios'. *Review of Economic Studies*, 86, 723–754.

Gruss, L. T. and Schmitt, D. (2015). 'The evolution of the human pelvis: changing adaptations to bipedalism, obstetrics and thermoregulation'. *Philosophical Transactions of the Royal Society B*, 370 (1663).

Guerra, F. (1977). 'The introduction of cinchona in the treatment of malaria'. Part I. *Journal of Tropical Medicine and Hygiene*, 80 (6), 112–118.

Gurven, M. and H. Kaplan, H. (2006). 'Determinants of time allocation across the lifespan'. *Human Nature*, 17, 1–49.

Gurven, M. and Kaplan, H. (2007). 'Longevity among hunter-gatherers: a cross-cultural examination'. *Population and Development Review*, 33 (2), 321–365.

Haidt, J. (2007). 'The new synthesis in moral psychology'. *Science*, 316, 998–1002.

Halpern, J. H., Pope, H. G., Sherwood, A. R., Barry, S., Hudson, J. I. and Yurgelun-Todd, D. (2004). 'Residual neuropsychological effects of illicit 3,4-methylenedioxymethamphetamine (MDMA) in individuals with minimal exposure to other drugs'. *Drug and Alcohol Dependence*, 75, 135–147.

Halpern, J. H., Sherwood, A. R., Hudson, J. I., Gruber, S., Kozin, D. and Pope Jr, H. G. (2010). 'Residual neurocognitive features of long-term ecstasy users with minimal exposure to other drugs'. *Addiction*, 106, 777–786.

Hames, G. (2014). *Alcohol in World History*. Routledge.

Han, F., Moughan, P. J., Li, J., Stroebinger, N. and Pang, S. (2021). 'The complementarity of amino acids in cooked pulse/cereal blends and effects on DIAAS'. *Plants*, 10 (10).

Hancock, L. E., Weiss, J. N., Duerr, G.M.E. (2010). 'Prospect Theory and the Framing of the Good Friday Agreement'. *Conflict Resolution Quarterly*, 28 (2), 183–203.

Hanes, W. T. and Sanello, F. (2002). *The Opium Wars: The Addiction of One Empire and the Corruption of Another*. Sourcebooks.

Hanlon, G. (2020). 'Historians and the Evolutionary Approach to Human Behaviour'. In: Workman, L., Reader, W. and Barkow, J. (ed.). *The Cambridge Handbook of Evolutionary Perspectives on Human Behaviour*. Cambridge University Press.

Hanson, T. (2015). *The Triumph of Seeds: how grains, nuts, kernels, pulses, and pips conquered the plant kingdom and shaped human history*. Basic Books.

Harbeck, M., Seifert, L., Hänsch, S., Wagner, D. M., Birdsell, D., Parise, K. L., Weichmann, I., Grupe, G., Thomas, A., Keim, P., Zöller, L., Bramanti, B., Riehm, J. M. and Scholz, H. C. (2013). '*Yersinia pestis* DNA from skeletal remains from the 6th century AD reveals insights into justinianic plague'. *PLOS Pathogens*, 9 (5).

Hareven, T. K. (1991). 'The history of the family and the complexity of social change'. *American Historical Review*, 96 (1), 95–124.

Harford, T. (2021). 'Cautionary tales – the curse of knowledge meets the Valley of Death'. Podcast. Available at: https://timharford.com/2021/04/cautionary-tales-the-charge-of-the-light-brigade/

Harper, K. (2015). 'Pandemics and passages to late antiquity: rethinking the plague of *c.*249–270 described by Cyprian'. *Journal of Roman Archaeology*, 28, 223–260.

Harper, K. (2017). *The Fate of Rome: Climate, Disease, and the End of an Empire*. Princeton University Press.

Harris, C. (2016). 'The Murder of Rasputin, 100 Years Later'. *Smithsonian Magazine*. Available at: https://www.smithsonianmag.com/history/murder-rasputin-100-years-later-180961572/

Harrison, H. (2017). 'The Quianlong emporer's letter to George III and the early-twentieth-century origins of ideas about traditional China's foreign relations'. *American Historical Review*, 122 (3), 680–701.

Harrison, M. (2013). *Contagion: how commerce has spread disease*. Yale University Press.

Hartung, J. (2010). 'Matrilineal inheritance: New theory and analysis'. *Behavioral and Brain Sciences*, 8 (4), 661–670.

Haselton, M. G., Nettle, D. and Andrews, P. W. (2015). 'The Evolution of Cognitive Bias'. In: Buss, D. M. (ed.). *The Handbook of Evolutionary Psychology*. Wiley.

Haskins, G. L. (1941). 'The beginnings of partible inheritance in the American colonies'. *Yale Law Journal*, 51, 1280–1315.

Hawkins, R. (1847). *The Observations of Sir Richard Hawkins, Knt, in his Voyage into the South Sea in the year 1593*. Hakluyt Society. Available at: https://www.gutenberg.org/cache/epub/57502/pg57502-images.html

He, W., Neil, S., Kulkarni, H., Wright, E., Agan, B. K., Marconi, V. C., Dolan, M. J., Weiss, R. A. and Ahuja, S. K. (2008). 'Duffy Antigen Receptor for Chemokines Mediates trans-Infection of HIV-1 from Red Blood Cells to Target Cells and Affects HIV-AIDS Susceptibility'. *Cell Host and Microbe*, 4 (1), 52–62.

Health and Human Services (2017). Press Release 26 October 2017: 'HHS Acting Secretary Declares Public Health Emergency to Address National Opioid Crisis'.

Helgason, A. Palsson, S.,Gudbjartsson, D. F., Krįstjansson, T. and

Stefansson, K. (2008). 'An association between the kinship and fertility of human couples'. *Science*, 319, 813–816.

Herlihy, D. (1997). *The Black Death and the transformation of the West*. Harvard University Press.

Herre, B. and Roser, M. (2013). 'Democracy. Our World in Data'. Available at: https://ourworldindata.org/democracy

Hess, S. (2015). *America's political dynasties: from Adams to Clinton*. Brookings Institution Press.

Hibbert, C. (2007). *Edward VII: the last Victorian king*. St Martin's Press.

Hibbert, C. (1961). *The Destruction of Lord Raglan*. Longman.

Hill, D. (2008). *1788: The Brutal Truth of the First Fleet*. William Heinemann: Australia.

Hill, K., Boesch, C., Goodall, J., Pusey, A., Williams, J. and Wrangham, R. (2001). 'Mortality rates among wild chimpanzees'. *Journal of Human Evolution*, 40, 437–50.

Ho, T.N.T, Abraham, N. and Lewis, R. J. (2020). 'Structure-function of neuronal nicotinic acetylcholine receptor inhibitors derived from natural toxins'. *Frontiers in Neuroscience*, 14.

Hodge, F. W. (1912). *Handbook of American Indians North of Mexico*. Smithsonian Institution: Bureau of American Ethnology. Bulletin 30.

Holden, B. A., Fricke, T. R., Wilson, D. A., Jong, M., Naidoo, K. S., Sankaridurg, P., Wong, T. Y., Naduvilath, T. J. and Resnikoff, S. (2016). 'Global prevalence of myopia and high myopia and temporal trends from 2000 through 2050'. *Ophthalmology*, 123 (5), 1036–1042.

Holden, C. J. (2002). 'Bantu language trees reflect the spread of farming across sub-Saharan Africa: a maximum-parsimony analysis'. *Proceedings of the Royal Society B*, 269 (1493).

Holmes, P. (2013). 'Tsetse-transmitted trypanosomes – their biology, disease impact and control'. *Journal of Invertebrate Pathology*, 112, Supplement 1, S11–S14.

Holtz, D. and Fradkin, A. (2020). 'Tit for tat? The difficulty of designing two-sided reputation systems'. *Sciendo*, 12 (2), 34–39.

Honigsbaum, M. (2020). *The pandemic century: a history of global contagion from the Spanish Flu to Covid-19*. W.H. Allen.

Hopkins, D. R. (2002). *The greatest serial killer: smallpox in history*. University of Chicago Press.

Howard, J. (2019). 'Gambler's Fallacy and Hot Hand Fallacy'. In: Howard, J., *Cognitive Errors and Diagnostic Mistakes*. Springer.

Hrdy, S. B. and Judge, D. S. (1993). 'Darwin and the puzzle of primogeniture'. *Human Nature*, 4, 1–45.

Huebner, S. R. (2021). 'The "Plague of Cyprian": A revised view of the origin and spread of a 3rd-century CE pandemic'. *Journal of Roman Archaeology*, 34(1), 1–24.

Hurley, T. D. (2012). 'Genes Encoding Enzymes Involved in Ethanol Metabolism'. *Alcohol Research*, 34 (3), 339–344.

Jaeggi, A. V., Gurven, M. (2013). 'Reciprocity explains food sharing in humans and other primates independent of kin selection and tolerated scrounging: A phylogenetic meta-analysis'. *Proceedings of the Royal Society* B, 280 (1768).

Jakobsson, M., Pearce, C., Cronin, T. M., Backman, J., Anderson, L. G., Barrientos, N., Björk, G., Coxall, H., de Boer, A., Mayer, L. A., Mörth, C.M., Nilsson, J., Rattray, J. E., Stranne, C., Semiletov, I. and O'Regan, M. (2017). 'Post-glacial flooding of the Bering Land Bridge dated to 11 cal ka BP based on new geophysical and sediment records'. *European Geosciences Union*, 13, 991–1005.

Jennings, J., Antrobus, K. L., Atencio, S. K., Glavich, E., Johnson, R., Loffler, G. and Luu, C. (2005). 'Drinking beer in a blissful mood: alcohol production, operational chains, and feasting in the ancient world'. *Current Anthropology*, 46 (2), 275–303.

Jensen, K., Call, J., Tomasello, M. (2007). 'Chimpanzees Are Vengeful But Not Spiteful'. *Proceedings of the National Academy of Sciences*, 104(32), 13046–50.

Johnson, D.D.P. and MacKay, N. J. (2015). 'Fight the power: Lanchester's laws of combat in human evolution'. *Evolution and Human Behavior*, 36, 152–163.

Johnson, R. J., Andrews, P., Benner, S. A. and Oliver, W. (2010). 'Theodore E. Woodward Award: The Evolution of Obesity: Insights from the mid-Miocene'. *Transactions of the American Clinical and Climatological Association*, 121, 295–308.

Jones, E. (2006). 'The Psychology of Killing: The Combat Experience of British Soldiers during the First World War'. *Journal of Contemporary History*, 41(2), 229–246.

Jones, O. D. (2015). 'Evolutionary Psychology and the Law'. In: Buss, D. M. (ed.). *The Handbook of Evolutionary Psychology*. Wiley.

Josefson, D. (1998). 'CF gene may protect against typhoid fever'. *British Medical Journal*, 316.

Kahneman, D. (2012). *Thinking, Fast and Slow*. Penguin.

Kahneman, D. and Tversky, A. (1979). 'Prospect Theory: An Analysis of Decision under Risk'. *Econometrica*, 47 (2), 263–291.

Kahneman, D., Knetsch, J. L. and Thaler, R. H. (1991). 'Anomalies: the endowment, effect, loss aversion, and status quo bias'. *Journal of Economic Perspectives*, 5 (1), 193–206.

Kalant, H. (1997). 'Opium revisited: a brief review of its nature, composition, non-medical use and relative risks'. *Addiction*, 92 (3), 267–277.

Kanakogi, Y., Miyazaki, M., Takahashi, H., Yamamoto, H., Kobayashi, T. and Hiraki, K. (2022). 'Third-party punishment by preverbal infants'. *Nature Human Behaviour*, 6, 1234–1242.

Kapoor, A. (2020). 'An unwanted shipment: The Indian experience of the 1918 Spanish flu'. *Economic Times*, 3 April 2020. Available at: https://economictimes.indiatimes.com/news/politics-and-nation/an-unwanted-shipment-the-indian-experience-of-the-1918-spanish-flu/articleshow/74963051.cms

Kato, G. J., Piel, F. B., Reid, C. D., Gaston, M. H., Ohene-Frempong, K., Krishnamurti, L., Smith, W. R., Panepinto, J. A., Weatherall, D. J., Costa, F. F. and Vichinsky, E. P. (2018). 'Sickle cell disease'. *Nature Reviews Disease Primers*, 4, article number: 18010.

Katz, S. H. and Voigt, M. M. (1986). 'Bread and beer: the early use of cereals in the human diet'. *Expedition*, 28 (2),23–34.

Kedishvili, N. Y. (2017). 'Retinoic acid synthesis and degradation'. *Subcellular Biochemistry*, 81, 127–161.

Kendrick, K.M. (2005). 'The neurobiology of social bonds'. *Journal of Neuroendocrinology*, 16 (12), 1007–1008.

Kenneally, C. (2014). *The invisible History of the Human Race: how DNA and history shape our identities and our futures*. Penguin.

Kesternich, I., Siflinger, B., Smith, J. P., Steckenleiter, C. (2020). 'Unbalanced sex ratios in Germany caused by World War II and their effect on fertility: A life cycle perspective'. *European Economic Review*, 30, 103581.

Khateeb, J., Li, Y. and Zhang, H. (2021). 'Emerging SARS-CoV-2 variants of concern and potential intervention approaches'. *Critical Care*, 25, 244.

Kimball, D. (2022). 'U.S.-Russian nuclear arms control agreements at a glance'. Arms Control Association. Available at: https://www.armscontrol.org/factsheets/USRussiaNuclearAgreements

King, M. W. (2019). 'How brain biases prevent climate action'. BBC Future. Available at: https://www.bbc.com/future/article/20190304-human-evolution-means-we-can-tackle-climate-change

Kingsbury, J. M. (1992). 'Christopher Columbus as a botanist'. *Arnoldia*, 52 (2), 11–28.

Klein, G., Felovich, P. J., Bradshaw, J. M. and Woods, D. D. (2005).' Common ground and coordination in joint activity'. In: Rouse, W. B. and Boff, K. R. (eds). *Organizational Simulation*. Wiley.

Kluesner, N. H. and Miller, D. G. (2014). 'Scurvy: Malnourishment in the Land of Plenty'. *Journal of Emergency Medicine*, 46 (4), 530–532.

Knobloch-Westerwick, S., Liu, L., Hino, A., Westerwick, A. and Johnson, B. K. (2019). 'Context impacts on confirmation bias: evidence from the 2017 Japanese snap election compared with American and German findings'. *Human Communication Research*, 45 (4), 427–449.

Knobloch-Westerwick, S., Mothes, C. and Polavin, N. (2017). 'Confirmation bias, ingroup bias and negativity bias in selective exposure to political information'. *Communication Research*, 47 (1).

Knobloch-Westerwick, S., Mothes, C., Johnson, B. K., Westerwick, A. and Donsbach, W. (2015). 'Political online information searching in Germany and the United States: confirmation bias, source credibility and attitude impacts'. *Journal of Communication*, 65 (3), 489–511.

Knowles, E. (2005). 'Providence is always on the side of the big battalions'. In *The Oxford Dictionary of Phrase and Fable*. Oxford University Press.

Koch, A., Brierley, C., Maslin, M. A. and Lewis, S. L. (2019). 'Earth system impacts of the European arrival and Great Dying in the Americas after 1492'. *Quaternary Science Reviews*, 207, 13–36.

Koehler, D. K. and Harvey, N. (eds) (2004). *Blackwell Handbook of Judgement and Decision Making*. Wiley.

Kokkonen, A. and Sundell, A. (2017). 'The King is Dead: political succession and war in Europe, 1000–1799'. Working Papers 2017:9. University of Gothenburg.

Kolata, G. (2001). *Flu: the story of the great influenza pandemic of 1918 and the search for the virus that caused it*. Atria Books.

Kramer, K. L. (2019). 'How there got to be so many of us: the evolutionary story of population growth and a life history of cooperation'. *Journal of Anthropological Research*, 45 (4).

Kramer, S. (2020). 'Polygamy is rare around the world and mostly confined to a few regions'. Pew Research Centre. Available at: https://www.pewresearch.org/fact-tank/2020/12/07/polygamy-is-rare-around-the-world-and-mostly-confined-to-a-few-regions/

Krebs, D. (2015). 'The Evolution of Morality'. In: Buss, D. M. (ed.). *The Handbook of Evolutionary Psychology*. Wiley.

Kringelbach, M. L., Phil, D. and Berridge, K. C. (2010). 'The functional neuroanatomy of pleasure and happiness'. *Discover Medicine*, 9 (49), 579–587.

Kruska, D.C.T (2014). 'Comparative quantitative investigations on brains of wild cavies (*Cavia aperea*) and guinea pigs (*Cavia aperea f. porcellus*). A contribution to size changes of CNS structures due to domestication'. *Mammalian Biology*, 79, 230–239.

Kuitems, M., Wallace, B. L., Lindsay, C., Scifo, A., Doeve, P., Jenkins, K., Lindauer, S., Erdil, P., Ledger, P. M., Forbes, V., Vermeeren, C., Friedrich, R. and Dee, M. W. (2021). 'Evidence for European presence in the Americas in AD 1021'. *Nature*, 601, 388–391.

Kurzban, R. and Neuberg, S. (2015). 'Managing Ingroup and Outgroup Relationships'. In: Buss, D. M. (ed.). *The Handbook of Evolutionary Psychology*. Wiley.

Lacey, K. and Lennon, J. T. (2016). 'Scaling laws predict global microbial diversity'. *Proceedings of the National Academy of Sciences of the United States of America*, 113 (21), 5970–5975.

Lalani, A. S., Masters, J., Zeng, W., Barrett, J., Pannu, R. and Everett, H. (1999). 'Use of chemokine receptors by poxviruses'. *Science*, 286 (5446), 1968–71.

Lamb, J. (2001). *Preserving the Self in the South Seas 1680–1840*. University of Chicago Press.

Landes, D. S. (2004). *Dynasties: Fortunes and Misfortunes of the World's Great Family Businesses*. Viking.

Lee, H. J., Macbeth, A. H., Pagani, J. H. and Scott Young, W. (2009). 'Oxytocin: the great facilitator of life'. *Progress in Neurobiology*, 88 (2), 127–151.

Leeson, P. T. (2007). 'An-arrgh-chy: the law and economics of pirate organization'. *Journal of Political Economy*, 115, 1049–1094.

Lents, N. (2018). *Human Errors: a panorama of our glitches, from pointless bones to broken genes*. Weidenfeld & Nicolson.

Lerman, A. E. and Acland, D. (2018). 'United in states of dissatisfaction: confirmation bias across the partisan divide'. *American Politics Research*, 48 (2).

Levy, B. (2009). *Conquistador: Hernan Cortes, King Montezuma, and the last stand of the Aztecs*. Bantam Books Inc.

Leyman, C. S. (1986). 'A review of the technical development of Concorde'. *Progress in Aerospace Sciences*, 23, 185–238.

Li, S., Schlebusch, C. and Jakobsson, M. (2014). 'Genetic variation reveals large-scale population expansion and migration during the expansion of Bantu-speaking peoples'. *Proceedings of the Royal Society B*, 281 (1793).

Liberman, P. (2001). 'The rise and fall of the South African bomb'. *International Security*, 26 (2), 45–86.

Lichtsinn, H. S., Weyand, A. C., McKinney, Z. J. and Wilson, A. M. (2021). 'Sickle cell trait: an unsound cause of death'. *Lancet*, 398 (10306), 1128–1129.

Lindenfors, P., Wartel, A. and Lind, J. (2021). 'Dunbar's number deconstructed'. *Biology Letters*, 17 (5).

Linster, C. L. and Van Schaftingen, E. (2007). 'Vitamin C biosynthesis, recycling and degradation in mammals'. *FEBS Journal*, 274, 1–22.

Lisuma, J., Mbega, E. and Ndakidemi, P. (2020). 'Influence of tobacco plant on macronutrient levels in sandy soils'. *MDPI*, 10 (3), 418.

Little, L. K. (2006). 'Life and Afterlife of the First Plague Pandemic'. In: Little, L. K. (ed.). *Plague and the End of Antiquity: The Pandemic of 541–750*. Cambridge University Press.

Livneh, Y. (2019). 'Overcoming the Loss Aversion Obstacle in Negotiation'. *Harvard Negotiation Law Review*, 25, 187–212.

Lloyd, C. C. (1981) 'Victualling of the fleet in the eighteenth and nineteenth centuries'. In: Watt, J., Freeman, E. J. and Bynum, W. F., (eds.) *Starving Sailors: The Influence of Nutrition upon Naval and Maritime History*. National Maritime Museum.

Loades, D. (2003). *Elizabeth I: The Golden Reign of Gloriana*. Bloomsbury.

Lovejoy, C. O. (1981). 'The origin of man'. *Science*, 211 (4480), 341–350.

Lovejoy, P. (1989). 'The Impact of the Atlantic Slave Trade on Africa: A Review of the Literature'. *The Journal of African History*, 30(3), 365–94

Lovejoy, P. (2000). *Transformations in Slavery: A History of Slavery in Africa*, 2nd ed. Cambridge University Press.

Lu Yu (c.760). 'The Classic of Tea'. Available as a translation in: Carpenter, F. R. (1974) *The Classic of Tea: Origins & Rituals*. The Ecco Press.

Luca, M. (2016). 'Designing online marketplaces: trust and reputation mechanisms'. *Innovation Policy and the Economy*, 17.

Luttinger, N. (2006). *The Coffee Book: Anatomy of an Industry from Crop to the Last Drop*. The New Press.

Mabbett, T. (2005). 'Tobacco nutrition and fertiliser use'. *Tobacco Journal International*, 6, 62–66.

Macdonald, J. (2014). *Feeding Nelson's Navy: the true story of food at sea in the Georgian era*. Frontline Books.

MacDonald, K. (1995). 'The establishment and maintenance of socially imposed monogamy in Western Europe'. *Politics and the Life Sciences*, 14 (1), 3–23.

Mackowiak, P. A., Blos, V. T., Aguilar, M. and Buikstra, J. E. (2005). 'On the origin of American Tuberculosis'. *Clinical Infectious Diseases*, 41, 515–518.

Maddieson, I. (1984). *Patterns of Sounds*. Cambridge University Press.

Mahan, A. T. (1895). 'Blockade in Relation to Naval Strategy'. *U.S. Naval Institute Proceedings*, 21(4), 76.

Majander, K., Pfrengle, S., Kocher, A., Neukamm, J., du Plessis, L., Pla-Díaz, M., Arora, N., Akgül, G., Salo, K., Schats, R., Inskip, S., Oinonen, M., Valk, H., Malve, M., Kriiska, A., Onkamo, P., González-Candelas, F., Kühnert, D., Krause, J. and Scheunemann, V. J. (2020). 'Ancient Bacterial Genomes Reveal a High Diversity of *Treponema pallidum* Strains in Early Modern Europe'. *Current Biology*, 30 (19), 3788–3803.

Malaney P.I.A., Spielman, A., Sachs, J. (2004). 'The Malaria Gap'. In: Breman, J. G., Alilio, M. S. and Mills, A., (eds). *The Intolerable Burden of Malaria II: What's New, What's Needed*: Supplement to Volume 71 (2) of the *American Journal of Tropical Medicine and Hygiene*. Northbrook (IL): American Society of Tropical Medicine and Hygiene.

Mann, C. C. (2011). *1493: Uncovering the New World Columbus Created*. Vintage.

Manning, P. (1990). *Slavery and African Life: Occidental, Oriental, and African Slave Trades*. Cambridge University Press.

Marcus, G. (2008). *Kluge: The Haphazard Evolution of the Human Mind*. Faber & Faber.

Marklein, K. E., Torres-Rouff, C., King, L. M. and Hubbe, M. (2019). 'The Precarious State of Subsistence: Reevaluating Dental Pathological Lesions Associated with Agricultural and Hunter-Gatherer Lifeways'. *Current Anthropology*, 60 (3), 341–368.

Marks, R. B. (2012). 'The (Modern) World since 1500'. In: McNeill, J. R. and Mauldin, E. S. (eds.). *A Companion to Global Environmental History*. Wiley.

Marr, A. (2013). *A History of the World*. Pan.

Martin, D. L. and Goodman, A. H. (2002). 'Health conditions before Columbus: paleopathology of native North Americans'. *Western Journal of Medicine*, 176 (1), 65–68.

Martin, S. (2015). *A Short History of Disease: from the Black Death to Ebola*. No Exit Press.

Massen, J.J.M., Ritter, C., Bugnyar, T. (2015). 'Tolerance and reward equity predict cooperation in ravens (*Corvus corax*)'. *Scientific Reports*, 5, 15021.

Massie, R. K. (1989). *Nicholas and Alexandra*. Victor Gollancz.

Mattison, S. M., Smith, E. A., Shenk, M. K. and Cochrane, E. E. (2016). 'The evolution of inequality'. *Evolutionary Anthropology*, 25, 184–199.

McCandless, P. (2007). 'Revolutionary fever: disease and war in the Lower South, 1776–1783'. *Transactions of the American Clinical and Climatological Association*, 118, 225–249.

McCarty, C., Killworth, P. D., Russell Bernard, H., Johnsen, E. C. and Shelley, G. A. (2001). 'Comparing two methods for estimating network size'. *Human Organization*, 60 (1), 28–39.

McCord, C. P. (1971). 'Scurvy as an occupational disease'. *Journal of Occupational Medicine*, 13 (6), 306–307.

McDermott, R. (2004). 'Prospect theory in political science: gains and losses from the first decade'. *Political Psychology*, 25 (2), 289–312.

McDermott, R. (2009). 'Prospect Theory and Negotiation'. In: Sjöstedt, G., Avenhaus, R. (eds). *Negotiated Risks: International Talks on Hazardous Issues*. Springer.

McEvedy, C. and Jones, R. (1977). *Atlas of World Population History*. Penguin.

McGee, H. (2004). *McGee on Food and Cooking: an encyclopedia of kitchen science, history and culture*. Hodder & Stoughton.

McGovern, P. E. (2018). *Ancient Brews: Rediscovered and Re-created*. W. W. Norton & Company.

McNeill, J. R. (2010). *Mosquito Empires: ecology and war in the Greater Caribbean, 1620–1914.* Cambridge University Press.

McNeill, W. H. (1976). *Plagues and Peoples.* Anchor Press.

Meekers, D. and Franklin, N. (1995). 'Women's perceptions of polygyny among the Kaguru of Tanzania'. *Ethnology*, 34 (4), 315–329.

Meletis, J. and Konstantopoulos, K. (2004). 'Favism – From the "avoid fava beans" of Pythagoras to the present'. *Haema*, 7 (1), 17–21.

Mercer, J. (2005). 'Prospect theory and political science'. *Annual Review of Political Science*, 8, 1–21.

Miller, G. A. (1956). 'The magical number seven, plus or minus two: Some limits on our capacity for processing information'. *Psychological Review*, 63 (2), 81–97.

Miller, K. M. (2016). *The Darien Scheme: Debunking the Myth of Scotland's Ill-Fated American Colonization Attempt.* Wright State University.

Milov, S. (2019). *The Cigarette: A Political History.* Harvard University Press.

Mineur, Y. S., Abizaid, A., Rao, Y., Salas, R., Dileone, R. J., Gündisch, D., Di-Ano, S., De Biasi, M., Horvath, T. L., Gao, X. B. and Picciotto, M. R. (2011). 'Nicotine decreases food intake through activation of POMC neurons'. *Science*, 332 (6035), 1330–1332.

Miron, J. A. and Zwiebel, J. (1991). *Alcohol Consumption During Prohibition.* National Bureau of Economic Research.

Mishra, S. and Mishra, M. B. (2013). 'Tobacco: its historical, cultural, oral and peridontal health association'. *Journal of International Society of Preventive and Community Dentistry*, 3 (1), 12–18.

Mitani, J. C., Watts, D. P. and Amsler, S. J. (2010). 'Lethal intergroup aggression leads to territorial expansion in wild chimpanzees'. *Current Biology*, 20 (2), R508–R508.

Mitchell, B. L. (2018). 'Sickle cell trait and sudden death'. *Sports Medicine – Open*, 4 (19).

Mitchell, S. (2006). *A History of the Later Roman Empire, AD 284 – AD 641: The Transformation of the Ancient World.* Wiley-Blackwell.

Mohandas, N. and An, X. (2012). 'Malaria and Human Red Blood Cells'. *Medical Microbiology and Immunology*, 201 (4), 593–598.

Monaghan, J. and Just, P. (2000). *Social and Cultural Anthropology: A Very Short Introduction.* Oxford University Press.

Montesquieu (1777). *The Spirit of the Laws*, Book XXVI, Chapter XVI. Text available at: https://oll.libertyfund.org/title/montesquieu-complete-works-4-vols-1777

Monot, M., Honoré, N., Garnier, T., Araoz, R., Coppée, J. Y., Lacroix, C., Sow, S., Spencer, J. S., Truman, R. W., Williams, D., Gelber, R., Virmond, M., Flageul, B., Cho, S. N., Ji, B., Paniz-Mondolfi, A., Convit, J., Young, S., Fine, P. E., Rasolofo, V., Brennan, P. J. and Cole, S. T. (2005). 'On the origin of leprosy'. *Science*, 308 (5724), 1040–1042.

Morland, P. (2019). *The Human Tide: How Population Shaped the Modern World*. John Murray Publishers.

Morland, P. (2022). 'Sinn Féin won the demographic war'. *UnHerd*, 10 May 2022. Available at: https://unherd.com/2022/05/sinn-fein-won-the-demographic-war/

Morris, I. (2011). *Why the West Rules – For Now: The Patterns of History and What They Reveal About the Future*. Profile Books.

Morris, I. (2014). *War: What is it Good For? The Role of Conflict in Civilisation, from Primates to Robots*. Profile Books.

Moser, D., Steiglechner, P. and Schlueter, A. (2021). 'Facing global environmental change: The role of culturally embedded cognitive biases'. *Environmental Development*, 44, 100735.

Mühlemann, B., Vinner, L., Margaryan, A., Wilhelmson, H., Castro, C., Allentoft, M. E., Damgaard, P., Hansen, A. J., Nielsen, S. H., Strand, L. M., Bill, J., Buzhilova, A., Pushkina, T., Falys, C., Khartanovich, V., Moiseyev, V., Jørkov, M.L.S., Sørensen, P. Ø., Magnusson, Y., Gustin, I., Schroeder, H., Sutter, G., Smith, G. L., Drosten, C., Fouchier, R.A.M., Smith, D. J., Willerslev, E., Jones, T. C. and Sikora, M. (2020) 'Diverse variola virus (smallpox) strains were widespread in northern Europe in the Viking Age'. *Science*, 369 (6502).

Münchau, W. (2017). 'From Brexit to fake trade deals – the curse of confirmation bias'. *Financial Times*, 9 July 2017. Available at: https://www.ft.com/content/b7d68798-62fb-11e7-91a7-502f7ee26895

Murdock, G. (1962). 'Ethnographic Atlas'. *Ethnology*, 1 (1), 113–134.

Nachman, M. W. and Crowell, S. L. (2000). 'Estimate of the mutation rate per nucleotide in humans'. *Genetics*, 156 (1), 297–304.

Nambi, K. (2020). 'How Spanish Flu brought independence to a country'. Available at: https://medium.com/lessons-from-history/how-spanish-flu-got-independence-to-a-country-f8d3f8fa6092

Nathanson, J. A. (1984). 'Caffeine and related methylxanthines: possible naturally occurring pesticides'. *Science*, 226 (4671), 184–187.

National Safety Council (2022). 'Injury Facts: Deaths in Public Places'. Available at: https://injuryfacts.nsc.org/home-and-community/deaths-in-public-places/introduction/

Newman, R. K. (1995). 'Opium smoking in late imperial China: a reconsideration'. *Modern Asian Studies*, 29 (4), 765–794.

Nickerson, R. S. (1998). 'Confirmation Bias: A Ubiquitous Phenomenon in Many Guises'. *Review of General Psychology*, 2(2), 175–220.

Nietzsche, F. (1888). *The Twilight of the Idols.* Translated by Anthony M. Ludovici (1911). Available at: https://www.gutenberg.org/files/52263/52263-h/52263-h.htm

Nishikimi, M., Kawai, T. and Yagi, K. (1992). 'Guinea pigs possess a highly mutated gene for L-Gulono-Y-lactone oxidase, the key enzyme for L-ascorbic acid biosynthesis missing in this species'. *Journal of Biological Chemistry*, 267 (30), 21967–21972.

Norn, S., Kruse, P. R. and Kruse, E. (2005). 'History of opium poppy and morphine'. *Dan Medicinhist Arbog*, 33, 171–184.

North, D. C. and Thomas, R. P. (1970). 'An Economic Theory of the Growth of the Western World'. *Economic History Review*, 23 (1), 1–17.

Novembre, J., Galvani, A. P., Slatkin, M. (2005). 'The geographic spread of the CCR5 Delta32 HIV-resistance allele'. *PLoS Biology*, 3 (11).

Nowak, M. A. (2006). 'Five rules for the evolution of cooperation'. *Science*, 314 (5805), 1560–1563.

Nowak, M. A. and Sigmund, K. (2005). 'Evolution of indirect reciprocity'. *Nature*, 437, 1291–1298.

Noymer, A. and Garenne, M. (2009). 'The 1918 influenza epidemic's effects on sex differentials in mortality in the United States'. *Population and Development Review*, 26 (3), 565–581.

Nunn, N. (2010). 'Shackled to the Past: The Causes and Consequences of Africa's Slave Trade'. In: Diamond, J. and Robinson, J. A. (eds). *Natural Experiments of History*. Harvard University Press.

Nunn, N. (2008). 'The Long-Term Effects of Africa's Slave Trades'. *Quarterly Journal of Economics*, 123, 139–176.

Nunn, N. (2017). 'Understanding the long-run effects of Africa's slave trades'. In: Michalopoulos, S. and Papaioannou, E. (eds). *The Long*

Economic and Political Shadow of History, Volume 2: Africa and Asia. CEPR Press.

Nunn, N. and Qian, N. (2010). 'The Columbian Exchange: a history of disease, food, and ideas'. *Journal of Economic Perspectives*, 24 (2), 163–188.

Nunn, N, and Wantchekon, L. (2011). 'The Slave Trade and the Origins of Mistrust in Africa'. *American Economic Review*, 101 (7), 3221–52.

O'Grady, M. (2020). 'What can we learn from the art of pandemics past?' *New York Times Style Magazine*. Available at: https://www.nytimes.com/2020/04/08/t-magazine/art-coronavirus.html

Öberg, M., Jaakkola, M., Woodward, A., Peruga, A. and Prüss-Ustün, A. (2011). 'Worldwide burden of disease from exposure to second-hand smoke: a retrospective analysis of data from 192 countries'. *Lancet*, 377 (9760), 8–14.

Ofcom (2019). 'Half of people now get their news from social media'. Ofcom, 24 July 2019. Available at: https://www.ofcom.org.uk/about-ofcom/latest/features-and-news/half-of-people-get-news-from-social-media

Office of Population Research (1946). 'War, Migration, and the Demographic Decline of France'. *Population Index*, 12 (2), 73–81. https://doi.org/10.2307/2730069

Ohler, N. (2016). *Blitzed: drugs in Nazi Germany*. Allen Lane.

Ojeda-Thies, C. and Rodriguez-Merchan, E. C. (2003). 'Historical and political implications of haemophilia in the Spanish royal family'. *Haemophilia*, 9 (2), 153–156.

Olds, J. and Peter, M. (1954). 'Positive reinforcement produced by electrical stimulation of septal area and other regions of rat brain'. *Journal of Comparative and Physiological Psychology*, 47 (6), 419–427.

Oldstone, M.B.A. (2009). *Viruses, Plagues, and History: Past, Present and Future*. Oxford University Press.

Ord, T. (2021). *The Precipice: Existential risk and the future of humanity*. Bloomsbury.

Organsk, A.F.K. (1958). World Politics. Alfred A. Knopf, New York. Ch. 5, p.132.

Ostlund, S. B. and Halbout, B. (2017). 'Mesolimbic Dopamine Signalling in Cocaine Addiction'. In: Preedy, V. R. (ed.). *The Neuroscience of Cocaine: Mechanisms and Treatment*. Academic Press.

Outram, Q. (2001). 'The socio-economic relations of warfare and the military mortality crises of the thirty years war'. *Medical History*, 45 (2), 151–184.

Owens, M. (2021). 'Afghanistan and the sunk cost fallacy'. *Washington Examiner*, 4 March 2021. Available at: https://www.washingtonexaminer.com/politics/afghanistan-and-the-sunk-cost-fallacy

Oxford, J. S. and Gill, D. (2018). 'Unanswered questions about the 1918 influenza pandemic: origin, pathology, and the virus itself'. *Lancet*, 18 (11), e348–e354.

Oyuela-Caycedo, A. and Kawa, N. C. (2015). 'A Deep History of Tobacco in Lowland South America'. In: Russell, A. and Rahman, E. (eds). *The Master Plant: Tobacco in Lowland South America*. Routledge.

Paine, L. (2015). *The Sea and Civilization: A Maritime History of the World*. Atlantic Books.

Pakendorf, B., Bostoen, K. and de Filippo, C. (2011). 'Molecular perspectives on the Bantu Expansion: a synthesis'. *Language Dynamics and Change*, 1, 50–88.

Pamuk, S. (2007). 'The Black Death and the origins of the "Great Divergence" across Europe, 1300–1600'. *European Review of Economic History*, 11 (3), 289–317.

Pamuk, S. and Shatzmiller, M. (2014). 'Plagues, Wages, and Economic Change in the Islamic Middle East, 700–1500'. *Journal of Economic History*, 74 (1), 196–229.

Pan, C. W., Ramamurthy, D. and Saw, S. M. (2011). 'Worldwide prevalence and risk factors for myopia'. 32 (1), 3–16.

Parker, G. (2008). Crisis and catastrophe: the global crisis of the seventeenth century reconsidered'. *American Historical Review*, 113 (4), 1053–1079.

Parker, G. (2020). *Emperor: A New Life of Charles V*. Yale University Press.

Parsons, R. (1996). 'The Mystery Bean'. *Los Angeles Times*, 18 April 1996. Available at: https://www.latimes.com/archives/la-xpm-1996-04-18-fo-59692-story.html

Payne, R. E. (2016). 'Sex, death, and aristocratic empire: Iranian jurisprudence in late antiquity'. *Comparative Studies in Society and History*, 58 (2), 519–549.

Pedersen, F. A. (1991). 'Secular trends in human sex ratios'. *Human Nature*, 2, 271–291.

Peirce, L.P. (1993). *The Imperial Harem: Women and Sovereignty in the Ottoman Empire*. Oxford University Press.

Pendergrast, M. (2009). *Coffee second only to oil? Is coffee really the second largest commodity?* Tea & Coffee Trade.

Pendergrast, M. (2010). *Uncommon Grounds: The History of Coffee and How it Transformed our World*. Basic Books.

Pereira, A. S., Kavanagh, E., Hobaiter, C., Slocombe, K. E. and Lameira, A. R. (2020). 'Chimpanzee lip-smacks confirm primate continuity for speech-rhythm evolution'. *Biology Letters*, 16 (5).

Perrin, F. (2022). 'On the origins of the demographic transition: rethinking the European marriage pattern'. *Cliometrica*, 16, 431–475.

Petrarch, F. (1348) Letter. Parma, Italy. Translation available in: Deaux, G. (1969) *The Black Death: 1347*. Weybright and Talley, New York.

Phillips, R. (2014). *Alcohol: a history*. UNC Press Books.

Phillips-Krawczak, C. (2014). 'Causes and consequences of migration to the Caribbean Islands and Central America: an evolutionary success story'. In: Crawford, M. H. and Campbell, B. C. (eds). *Causes and Consequences of Human Migration: An Evolutionary Perspective*. Cambridge University Press.

Pietraszewski, D. and Wertz, A. E. (2021). 'Why evolutionary psychology should abandon modularity'. *Perspectives on Psychological Science*, 17 (2).

Pinker, S. (2014). 'The Source of Bad Writing'. *Wall Street Journal*, 25 September 2104. Available at: https://www.wsj.com/articles/the-cause-of-bad-writing-1411660188

Pitre, M. C., Stark, R. J. and Gatto, M. C. (2016). 'First probable case of scurvy in ancient Egypt at Nag el-Quarmila, Aswan'. *International Journal of Paleopathology*, 13, 11–19.

Pittman, K. J., Glover, L. C., Wang, L. and Ko, D. C. (2016). 'The legacy of past pandemics: common human mutations that protect against infectious disease'. *PLOS Pathogens*, 12 (7).

Polansky, S. and Rieger, T. (2020). 'Cognitive biases: causes, effects and implications for effective messaging'. *NSI*. Available at: https://nsiteam.com/cognitive-biases-causes-effects-and-implications-for-effective-messaging/

Pollan, M. (2021). *This is Your Mind on Plants*. Penguin.

Poolman, E. M. and Galvani, A. P. (2006). 'Evaluating candidate agents of selective pressure for cystic fibrosis'. *Journal of the Royal Society Interface*, 4 (12).

Pope John XXIII. 'Pacem in Terris'. *The Holy See*, 11 April 1963, https://www.vatican.va/content/john-xxiii/en/encyclicals/documents/hf_j-xxiii_enc_11041963_pacem.html

Powers, S. T. and Lehmann, L. (2014). 'An evolutionary model explaining the Neolithic transition from egalitarianism to leadership and despotism'. *Proceedings of the Royal Society B*, 281.

Price, C. (2017). 'The Age of Scurvy'. *Science History Institute*. Available at: https://www.sciencehistory.org/distillations/magazine/the-age-of-scurvy

Price, D. T. (2014). 'New Approaches to the Study of the Viking Age Settlement across the North Atlantic'. *Journal of the North Atlantic*, 7, 1–12.

Rady, M. (2017). *The Habsburg Empire: A Very Short Introduction.* Oxford University Press.

Rady, M. (2020). *The Habsburgs: To Rule the World.* Basic Books.

Raihani, N. (2021). *The Social Instinct: How Cooperation Shaped the World.* Jonathan Cape.

Rajakumar , K. (2003). 'Vitamin D, Cod-Liver Oil, Sunlight, and Rickets: A Historical Perspective'. *Pediatrics*, 112 (2), e132–135.

Randy, E. E. (2010). 'Sickle cell trait in sports'. *Current Sports Medicine Reports*, 9 (6), 347–351.

Rao, P., Rodriguex, R. L. and Shoemaker, S. P. (2018). 'Addressing the sugar, salt, and fat issue the science of food way'. *NPJ Science of Food*, 2 (12).

Reich, D. (2018). *Who We Are and How We Got Here: Ancient DNA and the New Science of the Human Past.* Oxford University Press.

Riehn, R. K. (1990). *1812: Napoleon's Russian Campaign.* McGraw-Hill.

Ritzer, G. and Ryan, M. J. (2011). *The Concise Encyclopedia of Sociology.* Wiley.

Roberts, A. (2015). *Napoleon: a Life.* Penguin.

Roberts, J. M. and Westad, O. A. (2013). *The History of the World.* Oxford University Press.

Robson, S. L., van Schaik, C. P., Hawkes, K. (2006). 'The Derived Features of Human Life History'. In: Hawkes, K. and Paine, R. R.

The Evolution of Human Life History. School of American Research Press.

Roth, M. T. (1997). *Law Collections from Mesopotamia and Asia Minor*, 2nd edition. Society of Biblical Literature.

Rowold, D. J., Perez-Benedico, D., Stojkovic, O., Garcia-Bertrand, R., Herrera, R. J. (2016). 'On the Bantu expansion'. *Gene*, 593(1), 48-57.

Roxburgh, N. and Henke, J. S. (eds) (2020). *Psychopharmacology in British Literature and Culture, 1780–1900*. Palgrave Macmillan.

Rozenkrantz, L., D'Mello, A. M. and Gabrieli, J.D.E. (2021). 'Enhanced rationality in autism spectrum disorder'. *Trends in Cognitive Science*, 25 (8), 685–696.

Russell, B. (1950) *Unpopular Essays*. George Allen and Unwin, London

Rutherford, A. (2016). *A Brief History of Everyone Who Ever Lived: The Stories in Our Genes*. Weidenfeld & Nicolson.

Sapolsky, R. M. (2017). *Behave: The Biology of Humans at our Best and Worst*. Penguin.

Sarris, P. (2002). *The Justiniaic Plague: Origins and Effects*. Cambridge University Press.

Sarris, P. (2007). 'Bubonic Plague in Byzantium: The Evidence of Non-Literary Sources'. In: Little, L. K. (ed.). *Plague and the End of Antiquity: The Pandemic of 541–750*. Cambridge University Press.

Schacht, R. and Mudler, M. B. (2015). 'Sex ratio effects on reproductive strategies in humans'. *Royal Society Open Science*, 2 (1).

Schæbel, L. K., Bonefeld-Jørgensen, E. ., Laurberg, P., Vestergaard, H. and Andersen, S. (2015). 'Vitamin D-rich marine Inuit diet and markers of inflammation – a population-based survey in Greenland'. *Journal of Nutritional Science*, 4, 40.

Schaub, G. (2004). 'Deterrence, Compellence, and Prospect Theory'. *Political Psychology*, 25 (3), 389–411.

Scheidel, W. (1996). 'Brother-sister and parent-child marriage outside royal families in ancient Egypt and Iran: A challenge to the sociobiological view of incest avoidance?' *Ethnology and Sociobiology*, 17 (5), 319–340.

Scheidel, W. (2009a). 'A peculiar institution? Greco-Roman monogamy in global context'. *History of the Family*, 14, 280–291.

Scheidel, W. (2009b). *Sex and Empire: a Darwinian perspective*. Oxford University Press.

Schenck, T. (2019). *Holy Grounds: the surprising connection between coffee and faith – from dancing goats to Satan's drink*. Fortress Press.

Schino, G. and Aureli, F. (2010). 'Primate reciprocity and its cognitive requirements'. *Evolutionary Anthropology*, 19, 130–135.

Schmitt, D. P. (2015). 'Fundamentals of Human Mating Strategies'. In: Buss, D. M. (ed.). *The Handbook of Evolutionary Psychology*. Wiley.

Schudellari, M. (2021). 'How the coronavirus infects cells – and why Delta is so dangerous'. *Nature*, 595, 640–644.

Schwartz, B. (2006). 'The sunk-cost fallacy'. *Los Angeles Times*, 17 September 2006. Available at: https://www.latimes.com/archives/la-xpm-2006-sep-17-oe-schwartz17-story.html

Seebass, A. R. (1997). 'The Prospects for Commercial Supersonic Transport'. In: Sobieczky, H. (ed). *New Design Concepts for High Speed Air Transport*. Springer.

Ségurel, L. and Bon, C. (2017). 'On the evolution of lactase persistence in humans'. *Annual Review of Genomics and Human Genetics*, 12 (45).

Severin, T. (2008). *In Search of Robinson Crusoe*. Basic Books.

Shahraki, A. H., Carniel, E. and Mostafavi, E. (2016). 'Plague in Iran: its history and current status'. *Epidemology and Health*, 38.

Shammas, C. (1987). 'English inheritance law and its transfer to the colonies'. *American Journal of Legal History*, 31 (2), 145–163.

Sharp, P. M., Plenderleith, L. J. and Hahn, B. H. (2020). 'Ape Origins of Human Malaria'. *Annual Review of Microbiology*, 8 (74), 39–63.

Sherman, I. W. (2005). *The Power of Plagues*. ASM Press.

Sherman, I. W. (2007). *Twelve Diseases that Changed our World*. ASM Press.

Shermer, M. (2012). *The Believing Brain: From Spiritual Faiths to Political Convictions*. Robinson.

Simon, H. A. (1955). 'A behavioral model of rational choice'. *Quarterly Journal of Economics*, 69 (1), 99–118.

Simpson, T. (2012). *The Immigrants: The Great Migration from Britain to New Zealand, 1830–1890*. Penguin Random House New Zealand.

Singh, S. and Glowacki, L. (2022). 'Human social organization during the Late Pleistocene: Beyond the nomadic-egalitarian model'. *Evolution and Human Behavior*, 43 (5), 418–431.

Smith, D. S. (2003). 'Seasoning, Disease Environment, and Conditions of Exposure: New York Union Army Regiments and Soldiers'. In: Costa, D. L. (ed.). *Health and Labor Force Participation over the Life Cycle: Evidence from the Past*. University of Chicago Press.

Snelders, S. and Pieters, T. (2002). 'Speed in the Third Reich: methamphetamine (pervitin) use and a drug history from below'. *Social History of Medicine*, 24 (3), 686–699.

Sobolevskaya, O. (2013). 'The demographic echo of war'. *HSE University News*, 2 September 2013. Available at: https://iq.hse.ru/en/news/177669270.html

Solinas, M., Ferré, S., You, Z. B., Karcz-Kubicha, M., Popoli, P. and Goldberg, S. R. (2002). 'Caffeine induces dopamine and glutamate release in the shell of the nucleus accumbens'. *Brief Communication*, 22 (15), 6321–6324.

Southey, R. (1813). *The Life of Horatio Lord Nelson*. Available at: https://www.gutenberg.org/files/947/947-h/947-h.htm

Spiller, R. J. (1988). 'S.L.A. Marshall and the Ratio of Fire'. *RUSI Journal*, 133(4), 63–71.

Spinney, L. (2017). *Pale Rider: the Spanish flu of 1918 and how it changed the world*. Jonathan Cape.

Sporchia, F., Taherzadeh, O. and Caro, D. (2021). 'Stimulating environmental degradation: A global study of resource use in cocoa, coffee, tea and tobacco supply chains'. *Current Research In Environmental Sustainability*, 3, 1–11.

Standage, T. (2006). *A History of the World in 6 Glasses*. Bloomsbury.

Stanhope, A. (1840). *Spain under Charles the Second; or, Extracts from the correspondence of the Hon. Alexander Stanhope, British minister at Madrid, 1690–1699. From the originals at Chevening*. John Murray. Available at: https://wellcomecollection.org/works/xhq5ugzm

Stanovich, K. E., Toplak, M. E. and West, R. F. (2008). 'The development of rational thought: a taxonomy of heuristics and biases'. *Advances in Child Development and Behavior*, 36, 251–285.

Starkweather, K. E. and Hames, R. (2012). 'A survey of non-classical polyandry'. *Human Nature*, 23, 149–172.

Steele, J. (2002). 'Biological Constraints'. In: Hart, J. P. and Terrell, J. E. (eds). *Darwin and Archaeology: A Handbook of Key Concepts*. Bergin & Garvey.

Stephens, J. C., Reich, D. E., Goldstein, D. B., Shin, H. D., Smith, M. W., Carrington, M. (1998). 'Dating the origin of the CCR5-Delta32 AIDS-resistance allele by the coalescence of haplotypes'. *American Journal of Human Genetics*, 62(6), 1507–15.

Steppuhn, A., Gase, K., Krock, B., Halitschke, R. and Baldwin, I. T. (2004). 'Nicotine's Defensive Function in Nature'. *PLoS Climate*, 2 (10).

Stevens, R. (2005). 'The history of haemophilia in the royal families of Europe'. *British Journal of Haematology*, 105 (1), 25–32.

Stevenson, D. (2011). *With Our Backs to the Wall: Victory and Defeat in 1918*. Allen Lane.

Stevenson, P. C., Nicolson, S. W. Wright, G. A. (2017). 'Plant secondary metabolites in nectar: impacts on pollinators and ecological functions'. *Functional Ecology*, 31 (1), 65–75.

Stewart-Williams, S. (2018). *The Ape that Understood the Universe: How the Mind and Culture Evolve*. Cambridge University Press.

Stock, J. T. (2008). 'Are humans still evolving?' *Science and Society*, 9, 51–54.

Stone, A. C., Wilbur, A. K., Buikstra, J. E. and Roberts, C. A. (2009). 'Tuberculosis and leprosy in perspective'. *American Journal of Physical Anthropology*, 140 (S49), 66–94.

Stone, V. E., Cosmides, L., Tooby, J., Kroll, N. and Knight, R. T. (2002). 'Selective impairment of reasoning about social exchange in a patient with bilateral limbic system damage'. *Proceedings of the National Academy of Sciences of the United States of America*, 99 (17), 11531–11536.

Strachan, H. (2006). Training, Morale and Modern War'. *Journal of Contemporary History*, 41 (2), 211–227.

Strassmann, J. E. (1984). 'Female-Biased Sex Ratios in Social Insects Lacking Morphological Castes. *Evolution*, 38 (2), 256–266.

Surowiecki, J. (2004). *The Wisdom of Crowds: Why the Many Are Smarter Than the Few and How Collective Wisdom Shapes Business, Economies, Societies and Nations*. Doubleday.

Sussman, G. D. (2022). 'Was the Black Death in India and China?' *Bulletin of the History of Medicine*, 85 (3), 319–355.

Swallow, D. M. (2003). 'Genetics of lactase persistence and lactose intolerance'. *Annual Review of Genetics*, 37, 197–219.

Swerdlow, D. L., Mintz, E. D., Rodriguez, M., Tejada, E., Ocampo, C., Espejo, L. (1994). 'Severe life-threatening cholera associated with blood group O in Peru: implications for the Latin American epidemic'. *Journal of Infectious Diseases*, 170 (2), 468–72.

Talbot, J. E. (1991). 'Concorde development – powerplant installation and associated systems'. *SAE Transactions*, 100, 2681–2698.

Tana, V. D. and Hall, J. (2015). 'Isspresso development and operations'. *Journal of Space Safety Engineering*, 2 (1), 39–44.

Taubenberger, J. K. and Morens, D. M. (2006). '1918 influenza, the mother of all pandemics'. *Emerging Infectious Diseases*, 12 (1), 15–22.

Taylor, L. H., Latham, S. M. and Woolhouse, M.E.J. (2001). 'Risk factors for human disease emergence'. *Philosophical Transactions of the Royal Society B*, 356 (1411).

Teger, A. I. (1980). *Too Much Invested to Quit*. Pergamon.

Teso, E. (2019). 'The long-term effect of demographic shocks on the evolution of gender roles: evidence from the trans-Atlantic slave trade'. *Journal of the European Economic Association*, 17 (2), 497–534.

Tharoor, I. (2021). 'We're still living in the age of Napoleon'. *Washington Post*, 7 May 2021. Available at: https://www.washingtonpost.com/world/2021/05/07/napoleon-legacy-france/

The Boston Globe (2021). 'A sordid family affair'. *Boston Globe*. Available at: https://apps.bostonglobe.com/opinion/graphics/2021/06/future-proofing-the-presidency/part-3-a-sordid-family-affair/

The New York Times (1973). 'Supersonic civilian flights over U.S. are outlawed'. *New York Times*, 28 March 1973. Available at: https://www.nytimes.com/1973/03/28/archives/supersonic-civilian-flights-over-us-are-outlawed.html

The Royal Swedish Academy of Sciences (2002). Press Release: 'The Sveriges Riksbank Prize in Economic Sciences in Memory of Alfred Nobel 2002'. Available at: https://www.nobelprize.org/prizes/economic-sciences/2002/press-release/

The White House (2005). Report to the President, March 31, 2005. Available at: https://georgewbush-whitehouse.archives.gov/wmd/text/report.html

The White House (2017). Remarks by President Trump on the Strategy in Afghanistan and South Asia, 21 August 2017. Available at: https://trumpwhitehouse.archives.gov/briefings-statements/remarks-president-trump-strategy-afghanistan-south-asia/

Theofanopoulou, C., Gastaldon, S., O'Rourke, T., Samuels, B. D., Messner, A., Martins, P. T., Delogu, F., Alamri, S. and Boeckx, C. (2017). 'Self-domestication in Homo sapiens: insights from comparative genomics'. *PLOS ONE*, 13 (5).

Thornton, J. (1983).' Sexual Demography: The Impact of the Slave Trade on Family Structure'. In: Robertson, C. C. and Klein M. A. (eds). *Women and Slavery in Africa*, University of Winsconsin Press.

Thucydides, *The History of the Peloponnesian War*, Book II, Chapter VII. Translated by Richard Crawley (1874). Available at: Project Gutenberg: https://www.gutenberg.org/files/7142/7142-h/7142-h.htm

Tishkoff, S. A. Reed, F. A., Friedlaender, F. R., Ehret, C., Ranciaro, A., Froment, A., Hirbo, J. B., Awomoyi, A. A., Bodo, J. M., Doumbo, O., Ibrahim, M., Juma, A. T., Kotze, M. J., Lema, G., Moore, J. H., Mortensen, H., Nyambo, T. B., Omar, S. A., Powell, K., Pretorius, G. S., Smith, M. W., Thera, M. A., Wambebe, C., Weber, J. L. and Williams, S. M. (2009). 'The genetic structure and history of Africans and African Americans'. *Science*, 324 (5930), 1035–1044.

Tobin, V. (2009). 'Cognitive bias and the poetics of surprise'. *Language and Literature*, 18 (2), 155–172.

Toner, D. (2021). *Alcohol in the Age of Industry, Empire and War*. Bloomsbury Publishing.

Tooby, J. and Cosmides, L. (1996). 'Friendship and the Banker's Paradox: other pathways to the evolution of adaptations for altruism'. *Proceedings of the British Academy*, 88, 119–143.

Topik, S. (2004). 'The World Coffee Market in the Eighteenth and Nineteenth Centuries, from Colonial To National Regimes'. Working Paper No. 04/04. University of California, Irvine.

Trevathan, W.(2015). 'Primate pelvic anatomy and implications for birth'. *Philosophical Transactions of the Royal Society B*, 370 (1663).

Trivers, R. (2006). 'Reciprocal altruism: 30 years later'. In: Kappeler, P. M., van Schaik, C. P. (eds). *Cooperation in Primates and Humans*. Springer.

Trivers, R. L. (1971). 'The evolution of reciprocal altruism'. *Quarterly Review of Biology*, 46 (1), 35–57.

Tushingham, S., Ardura, D. A., Eerkens, J. W. and Palazoglu, M. (2013). 'Hunter-gatherer tobacco smoking: Earliest evidence from the Pacific Northwest Coast of North America'. *Journal of Archaeological Science*, 40 (2), 1397–1407.

Tversky, A. and Kahneman, D. (1974). 'Judgement under Uncertainty: Heuristics and Biases'. *Science*, 185 (4157), 1124–1131.

Tversky, A. and Kahneman, D. (1981). 'The framing of decisions and the psychology of choice'. *Science*, 311 (30), 453–458.

Unicef (2021). 'Vitamin A deficiency'. Available at: https://data.unicef.org/topic/nutrition/vitamin-a-deficiency/

United Nations Office on Drugs and Crime (2021). World Drug Report 2021. 'Drug Market Trends: Cannabis and Opioids'. United Nations.

United Nations Office on Drugs and Crime (2022). Afghanistan Opium Survey 2021. 'Cultivation and Production'. United Nations.

Vachula, R. S., Huang, Y., Longo, W. M., Dee, S. G., Daniels, W. C. and Russell, J. M. (2019). 'Evidence of ice age in humans in eastern Beringia suggests early migration to North America'. *Quaternary Science Reviews*, 205, 35–44.

Vale, B. (2008). 'The Conquest of Scurvy in the Royal Navy 1793–1800: A challenge to current orthodoxy'. *Mariner's Mirror*, 94(2), 160–75.

van Leengoed, E., Kerker, E. and Swanson, H. H. (1987). 'Inhibition of post-partum maternal behaviour in the rat by injecting an oxytocin antagonist into the cerebral ventricles'. *Journal of Endocrinology*, 112 (2), 275–282.

Verpoorte, R. (2005). 'Alkaloids'. *Encyclopedia of Analytical Science*, 2nd edition. Elsevier.

Vidmar, J. (2005). *The Catholic Church though the Ages: A History*. Paulist Press.

Vilas, R., Ceballos, F. C., Al-Soufi, L., González-García, R., Moreno, C., Moreno, M., Villanueva, L., Ruiz, L., Mateos, J., Gonzalez, D., Ruiz, J., Cinza, A., Monje, F. and Álvarez, G. (2019). 'Is the "Habsburg jaw" related to inbreeding?' *Annals of Human Biology*, 46 (7–8), 553–561.

Vis, B. (2011). 'Prospect theory and political decision making'. *Political Studies Review*, 9, 334–343.

Visceglia, M. A. (2002). 'Factions in the Sacred College in the sixteenth and seventeenth centuries'. In: Signorotto, G. and Visceglia, M. A. (eds). *Court and Politics in Papal Rome, 1492–1700*. Cambridge University Press.

Vishnevsky, A. and Shcherbakova, E. (2018). 'A new stage of demographic change: A warning for economists'. *Russian Journal of Economics*, 4 (3), 229–248.

Voelkl, B., Portugal, S. J., Unsöld, M., Usherwood, J. R., Wilson, A. M., Fritz, J. (2015). 'Matching times of leading and following suggest cooperation through direct reciprocity during V-formation flight in ibis'. *Proceedings of the National Academy of Sciences*, 112, 2115–2120.

Vogel, K. (1933). 'Scurvy – "The Plague of the Sea and the Spoyle of Mariners"'. *Bulletin of the New York Academy of Medicine*, IX (8).

Volkow, N. D. and Blanco, C. (2021). 'Research on substance use disorders during the COVID-19 pandemic'. *Journal of Substance Abuse Treatment*, 129, 1–3.

von Clausewitz, C. (1832). *On War*. Translated by Graham, J. J. (1874). Available at: Project Gutenberg https://www.gutenberg.org/ebooks/1946

Walker, M and Matsa, K. E. (2021). *News consumption across social media in 2021*. Pew Research Center. Available at: https://www.pewresearch.org/journalism/2021/09/20/news-consumption-across-social-media-in-2021/.

Walker, M. (2018). *Why We Sleep: The New Science of Sleep and Dreams*. Penguin.

Wallace, Birgitta. (2003). 'The Norse in Newfoundland: L'Anse aux Meadows and Vinland'. *Newfoundland Studies*, 19 (1), 5–43.

Walter, K. S., Carpi, G., Caccone, A. and Diuk-Wasser, M. A. (2017). 'Genomic insights into the ancient spread of Lyme disease across North America'. *Nature Ecology & Evolution*, 1, 1569–1576.

Walter, R. (1748). *A Voyage Round the World … by George Anson*. John and Paul Knapton, London. Reproduced in Household, H. W. (Ed.) (1901) *Anson's Voyage Around the World: The Text Reduced*. Rivingtons, London. Available at: https://www.gutenberg.org/files/16611/16611-h/16611-h.htm

Wason, P. C. (1968). 'Reasoning about rule'. *Quarterly Journal of Experimental Psychology*, 20 (3).

Wason, P. C. (1983). 'Realism and rationality in the selection task'. In: J. Evans (Ed.), *Thinking and reasoning: Psychological approaches*. Routledge.

Watson, A. (2015). *Ring of steel: Germany and Austria-Hungary at war, 1914–1918*. Penguin.

Watson, A. (2022). 'Social media as a news source worldwide 2022'. Statista. Available at: https://www.statista.com/statistics/718019/social-media-news-source/

Watson, P. (2012). *The Great Divide: History and Human Nature in the Old World and the New*. Weidenfeld & Nicolson.

Watt, J., Freeman, E. J. and Bynum, W. F. (1981). *Starving Sailors: influence of nutrition upon naval and maritime history*. National Maritime Museum.

Watts, S. (1999). *Epidemics and History: Disease, Power and Imperialism*. Yale University Press.

Weatherall, D. J. (2008). 'Genetic variation and susceptibility to infection: the red cell and malaria'. *British Journal of Haematology*, 141 (3), 276–286.

Webb, J.L.A. (2017). 'Early Malarial Infections and the First Epidemiological Transition'. In: Boivin, N., Crassard, R., Petraglia, M. (eds). *Human Dispersal and Species Movement: From Prehistory to the Present*. Cambridge University Press.

Webber, R. (2015). *Disease Selection: The Way Disease Changed the World*. CABI Publishing.

Wells, R. V. (1975). *Population of the British Colonies in America Before 1776: A Survey of Census Data*. Princeton University Press.

Whatley, C. (2001). *Bought and sold for English gold? The Union of 1707*. Tuckwell Press.

Whatley, W. and Gillezeau, R. (2011). 'The impact of the Transatlantic slave trade on ethnic stratification in Africa'. *American Economic Review*, 101 (3), 571–6.

Wheeler, B. J., Snoddy, A.M.E., Munns, C., Simm, P., Siafarikas, A. and Jefferies, C. (2019). 'A brief history of nutritional rickets'. *Frontiers in Endocrinology*, 10.

Wheelis M. (2002). 'Biological Warfare at the 1346 Siege of Caffa'. *Emerging Infectious Diseases*, 8 (9), 971–975.

White, A. (2020). 'Halle Berry says Pierce Brosnan saved her from choking during Bond sex scene gone wrong'. *Independent*. Available at: https://www.independent.co.uk/arts-entertainment/films/news/halle-berry-pierce-brosnan-james-bond-die-another-day-sex-scene-choking-a9477701.html

White, D. R., Betzig, L., Borgerhoff, M., Chick, G., Hartung, J., Irons, W., Low, B. S., Otterbein, K. F., Rosenblatt, P. C. and Spencer, P. (1988). 'Rethinking polygyny: co-wives, codes and cultural systems'. *Current Anthropology*, 29 (4), 529.

Wigner, E.P. (1960). 'The unreasonable effectiveness of mathematics in the natural sciences'. Richard Courant lecture in mathematical sciences delivered at New York University, May 11, 1959. *Communications on Pure and Applied Mathematics*, 13 (1), 1–14.

Wild, A. (2010). *Black Gold: the dark history of coffee*. Harper Perennial.

Wilke, A. and Clark Barrett, H. (2009). 'The hot hand phenomenon as a cognitive adaptation to clumped resources'. *Evolution and Human Behavior*, 30 (3), 161–169.

Willyard, C. (2018). 'New human gene tally reignites debate'. *Nature*, 558, 354–355

Williams, D. M. (1991). 'Mid-Victorian Attitudes to Seamen and Maritime Reform: The Society for Improving the Condition of Merchant Seamen, 1867'. *International Journal of Maritime History*, 3 (1),101–26.

Williams, T. N. (2011). 'How do hemoglobins S and C result in malaria protection?' *Journal of Infectious Diseases*, 204 (11), 1651–1653.

Wilson, D. M. (1989). *The Vikings and their Origins: Scandinavia in the first millennium*. Thames and Hudson.

Wilson, E. O. (2012) *The Social Conquest of Earth*. W. W. Norton & Co

Wilson, M. L., Boesch, C., Fruth, B., Furuichi, T., Gilby, I .C., Hashimoto, C., Hobaiter, C. L., Hohmann, G., Itoh, N., Koops, K., Lloyd, J. N., Matsuzawa, T., Mitani, J. C., Mjungu, D. C., Morgan, D., Muller, M. N., Mundry, R., Nakamura, M., Pruetz, J., Pusey, A. E., Riedel, J., Sanz, C., Schel, A. M., Simmons, N., Waller, M., Watts, D. P., White, F., Wittig, R. M., Zuberbühler, K. and Wrangham, R. W. (2014). 'Lethal aggression in Pan is better explained by adaptive strategies than human impacts'. *Nature*, 513, 414–417.

Winegard, T. (2019). *The Mosquito: A Human History of our Deadliest Predator*. Text Publishing Company.

Winkelman, M. J. and Sessa, B. (eds) (2019). *Advances in Psychedelic Medicine: state-of-the-art therapeutic applications*. Praeger.

Winston, R. (2003). *Human Instinct*. Bantam.

Wolf, M. (2012). 'The world's hunger for public goods'. *Financial Times*. Available at: http://www.ft.com/content/517e31c8-45bd-11e1-93f1-00144feabdco

Wolfe, N. D., Dunavan, C. P. and Diamond, J. (2007). 'Origins of major human infectious diseases'. *Nature*, 447, 279–283.

Wolfgang, E. (2006). *Man, Medicine, and the State: the human body as an object of government-sponsored medical research in the 20th century*. Digital Georgetown.

Wood, E. (1916). *Our Fighting Services*. Cassell.

Woodward, H. (2009). *A Brave Vessel: The True Tale of the Castaways who Rescued Jamestown*. Viking.

World Health Organisation (2021). WHO Report on the Global Tobacco Epidemic, 2021. Available at: https://www.who.int/publications/i/item/9789240032095

Wrangham, R. (2019). *The Goodness Paradox: How evolution made us both more and less violent*. Profile Books.

Wrangham, R. W. (1999). 'Evolution of coalitionary killing'. *American Journal of Physical Anthropology*, 110 (S29), 1–30.

Wright, G. A., Baker, D. D., Palmer, M. J., Stabler, D., Mustard, J. J., Power, E. F., Borland, A. M. and Stevenson, P. C. (2013). 'Caffeine in floral nectar enhances a pollinator's memory of reward'. *Science*, 339 (6124).

Wrigley, E. A. (1985). 'The fall of marital fertility in nineteenth-century France: Exemplar or exception?' (Part I). *European Journal of Population*, 1, 31–60.

Xue, Y., Wang, Q., Ng, B. L., Swerdlow, H., Burton, J., Skuce, C., Taylor, R., Abdellah, Z., Zhao, Y., MacArthur, D. G., Quail, M. A., Carter, N. P., Yang, H. and Tyler-Smith, C. (2009). 'Human Y chromosome base-substitution mutation rate measured by direct sequencing in a deep-rooting pedigree'. *Current Biology*, 19 (17), 1453–1457.

Xue, Y., Zerjal, T., Bao, W., Zhu, S., Lim, S. K., Shu, Q., Xu, J., Du. R., Fu, S., Yang, H. and Tyler-Smith, C. (2005). 'Recent spread of a Y-chromosomal lineage in Northern China and Mongolia'. *AJHG*, 77 (6), 1112–1116.

Yalcindag, E., Elguero, E., Arnathau, C., Durand, P., Akiana, J., et al. (2011). 'Multiple independent introductions of *Plasmodium falciparum* in South America'. *Proceedings of the National Academy of Sciences of the United States of America*, 109 (2), 511–516.

Yamagishi, T. (1986). 'The provision of a sanctioning system as a public good'. *Journal of Personality and Social Psychology*, 51 (1), 110–116.

Yasuoka, H. (2013). 'Dense wild yam patches established by hunter-gatherer camps: beyond the wild yam question, toward the historical ecology of rainforests'. *Human Ecology*, 41(1), 465–475.

Young, L. J. and Wang, Z. (2004). 'The neurobiology of pair bonding'. *Nature Neuroscience*, 7, 1048–1054.

Yu, Z. and Kibriya, S. (2016). 'The impact of the slave trade on current civil conflict in Sub-Saharan Africa'. Working paper, Texas A&M University.

Zabecki, D. T. (2001). *The German 1918 Offensives: a case study in the operational level of war*. Routledge.

Zahid, H. J., Robinson, E. and Kelly, R. L. (2015). 'Agriculture, population growth, and statistical analysis of the radiocarbon record'. *Proceedings of the National Academy of Sciences*, 113 (4), 931–935.

Zamoyski, A. (2019). 'The personality traits that led to Napoleon Bonaparte's epic downfall'. *History*. Available at: https://www.history.com/news/napoleon-bonaparte-downfall-reasons-personality-traits

Zaval, L. and Cornwell, J.F.M. (2016). 'Cognitive biases, non-rational judgements, and public perceptions of climate change'. In: *Oxford Research Encyclopedia of Climate Science*. Oxford University Press.

Zerjal, T., Xue, Y., Bertorelle, G., Wells, S., Bao, W., et al. (2003). 'The genetic legacy of the Mongols'. *American Journal of Human Genetics*, 72 (3), 717–721.

Zhang, Y, Xu, Z. P. and Kibriya, S. (2021). 'The long-term effects of the slave trade on political violence in Sub-Saharan Africa'. *Journal of Comparative Economics*, 49 (3), 776–800.

Zhao, J. and Luo, Y. (2021). 'A framework to address cognitive biases of climate change'. *Neuron*, 109 (22), 3548–51.

Zhao, T., Liu, S., Zhang, R., Zhao, Z., Yu, H., Pu, L., Wang, L. and Han, L. (2022). 'Global burden of vitamin A deficiency in 204 countries and territories from 1990–2019'. *Nutrients*, 14 (5), 950.

Zheng, Y. (2003). 'The social life of opium in China, 1483–1999'. *Modern Asian Studies*, 37 (1), 1–39.

Zhou, W. X., Sornette, D., Hill, R. A. and Dunbar, R.I.M. (2005). 'Discrete hierarchical organization of social group sizes'. *Proceedings of the Royal Society B*, 272 (1561), 439–444.

Zimmer, C. (2019). *She Has Her Mother's Laugh: The Story of Heredity, Its Past, Present and Future*. Picador.

Zimmerman, J. L. (2012). 'Cocaine intoxication'. *Critical Care Clinics*, 28 (4), 517–526.

Zinsser, H. (1935). *Rats, Lice and History*. Little, Brown.

Index

Acknowledgements

My first thank you must go to my agent Will Francis. Will is the only person to have been with each of my books from the very conception of a new project, when the idea is still inchoate and fuzzy-edged, through the development of the proposal, writing and to its birth on publication day, and his insightful guidance is a never-ending source of support. Huge thanks too to the rest of the outstanding team at Janklow and Nesbit in London: Kirsty Gordon, Mairi Friesen-Escandell, Ren Balcombe, Corissa Hollenbeck, Ellis Hazelgrove, Michael Steger, Maimy Suleiman, as well as PJ Mark and Ian Bonaparte in the New York office.

I'm also enormously grateful to Stuart Williams at The Bodley Head for so enthusiastically taking on this book for publication, and especially to Jörg Hensgen, whose keen editorial eye once again improved the manuscript and the clarity of my thinking immeasurably. He is the sculpting and polishing to my flabby first draft. Thanks too to Sam Wells and Fiona Brown for copy-editing and proofreading, respectively, Alex Bell for the index, and Rhiannon Roy and Laura Reeves for overseeing the whole publication process. I absolutely adore Kris Potter's eye-popping jacket design (who also worked his magic on the covers for *The Knowledge* and *Origins*). Incidentally, the human sketch is 'A Nude Throwing' by the eighteenth-century Swiss painter Henry Fuseli. Thanks also to Joe Pickering and

Carmella Lowkis for their help in the marketing and publicity of the finished book.

I've also benefitted enormously from the help so generously offered by various researchers and experts along the way, including (in alphabetical order): Koen Bostoen, Brad Elliot, Douglas Howard, Stephen Luscombe, Nichola Raihani, Liron Rozenkrantz, Alex Stewart, Kaj Tallungs, Yuhan Sohrab-Dinshaw Vevaina, and Amelia Walker – a huge thanks to each and every one of you. A special thank-you goes to my research assistants Rob Hampton, Sara Knudsen and Megan Bryant, who helped me wrangle my notes, chase down references, assemble the Endnotes and Bibliography, and a thousand and one other essential tasks.

Finally, and most heartfelt, is my gratitude to my wonderful wife Davina Bristow. She has been a pillar of unwavering support and encouragement, and this book wouldn't be what it is without her own scientific storyteller's eye. This book is dedicated to her, and our son Sebastian.